Air Pollution

W. Strauss and S. J. Mainwaring

Department of Industrial Science, University of Melbourne

Edward Arnold

© J. Strauss and S. J. Mainwaring 1984

First published 1984 by
Edward Arnold (Publishers) Ltd
41 Bedford Square, London WC1B 3DQ

Edward Arnold,
300 North Charles Street,
Baltimore, MD 21201, USA

Edward Arnold (Australia) Ltd.,
80 Waverley Road, Caulfield East,
Victoria 3145, Australia

British Library Cataloguing in Publication Data

Strauss, W.
  Air pollution.
  1. Air—Pollution
  I. Title   II. Mainwaring, S.J.
  363.7 '392   TD883

  ISBN 0-7131-3493-3

Printed in Great Britain by
Thomson Litho Ltd, East Kilbride, Scotland

Preface

The intention of this book was to make available some of the knowledge gained in years of specialized work in the field of air pollution control in a format accessible to non-specialists interested in the wide range of issues — scientific, industrial, economic, social and ecological — involved.

The text is suitable for later-year school students, for a variety of interdisciplinary tertiary courses and for general readers who wish to gain a substantial basic understanding of the concepts, the techniques and the developments of this very important twentieth century branch of applied science.

The work was considerably advanced when interrupted by the death of the original author Dr Werner Strauss, Reader in Chemical Engineering at the University of Melbourne. The text and final editing were completed by Dr Sylvia Mainwaring of the Department of Industrial Science.

Grateful acknowledgement is made to Mrs Angela Ghiggino for her patient and accurate work on the typescript.

SJM
1984

Contents

Preface   iii

1 What is Air Pollution?   1
Introduction   1
Air and Air Pollution   1
Dispersal of Air Pollution and the Weather   7

2 Sources of Air Pollution   12
Petroleum Refining   12
Smelting of Non-Ferrous Ores   15
Manufacture of Iron and Steel   18
Chemical Works and Other Industrial Processes   20
External Combustion Systems   25
Internal Combustion   29
Comparison of Pollution Sources   34

3 The Effects of Air Pollution   36
Air Pollution Episodes   38
Effects of Individual Pollutants on Man   47
Effects on Vegetation   55
Effects on Animals   60
Effects on Materials   62
The Cost   62
Global Changes   63

4 Measurement of Pollutants   71
Measurement of Source Concentrations   72

Measurement of Ambient Concentrations 75
Monitoring Methods 80
Calibration 85
Monitoring Networks 85

5 Air Pollution Control 87
Control of Gaseous Pollutants 89
Control of Particulate Pollutants 94
Disposal of Residual Air Pollutants – Chimney Stacks 102
Applications to Control Technology 105
Automobiles 108

6 Non-Technical Aspects of Control 112
Air Pollution Management 113
Implementation of Control Programmes 119

7 The Future 125
Domestic and Commercial Heating 125
Electric Power Generation 128
Transportation 139
Industrial Pollution 142

Bibliography 143

Index 146

1 What is Air Pollution?

Introduction

Air pollution is not really a new phenomenon, as smoke from heating and cooking fires, odours from domestic wastes – sewage and garbage – have been a characteristic of places where man has been living since grouping into communities began. However, in the past century in the developed countries, as methods of waste disposal, sewage treatment, domestic heating and cooking have changed, the traditional forms of air pollution – smoke and odours – have been reduced and replaced by a new set of air pollutants which result from our mobile, industrial, urban society. Overall, most of our air pollution problems today are a result of our industrial activities and our means of transportation; a result, in other words, of our use of energy.

Air and Air Pollution

Air is a mixture of gases which surrounds the earth in a comparatively thin layer. Most of the air (95%) is in the first 20 km above the earth's sea level, above which it decreases in density until it merges into the void of space some hundreds of kilometres above the earth. The lower part of this layer, the troposphere, is about 8 km thick at the earth's poles, and about twice this at the equator. Man's activities take place, for the most part, on the earth's surface within the first 2 km of the atmosphere. The pollutants produced by these activities are injected directly into the troposphere where they are mixed and transported.

The major constituents of air, nitrogen (78%), oxygen (20.94%) and argon (0.93%), do not react with one another under normal circumstances. Similarly, the trace components helium, neon, krypton, xenon, hydrogen and nitrous oxide have little or no interaction with other molecules. A number of other gases also present in trace quantities are not inert chemically but interact with the biosphere, the hydrosphere and each other, and so have a limited residence time in the atmosphere and characteristically variable concentrations (Table 1.1). It is this reactive group of gases which are considered pollutant when they are produced by man in sufficient quantities for the background concentrations of Table 1.1 to be significantly exceeded. The most important gases in this group are those which are universally present in the air of the world's cities,

Table 1.1   The composition of dry air in the lower troposphere (Free of water vapour)

	Chemical Symbol	Concentration[a] %	Calculated Residence Time
Principal Gases			
Nitrogen	N_2	73.0	Continuous
Oxygen	O_2	20.9	Continuous
Argon	A	0.93	Continuous
Carbon Dioxide	CO_2	0.032[b]	20 years[c]
Trace Gases			
(a) Permanent Gases *(non-reactive)*		ppm	
Helium	He	5.2	Continuous
Neon	Ne	18.0	Continuous
Krypton	Kr	1.1	Continuous
Xenon	Xe	0.086	Continuous
Hydrogen	H_2	0.5	?
Nitrous Oxide	N_2O	0.25	8–10 years
(b) *Reactive Gases*			
Carbon Monoxide	CO	0.1	0.2–0.3 years
Methane	CH_4	1.4	< 2 years
Non-Methane Hydrocarbons	'HC'	0.02	?
Nitric Oxide	NO	0.2 to 2.0×10^{-3}	2–8 days
Nitrogen Dioxide	NO_2	0.5 to 4.0×10^{-3}	
Ammonia	NH_3	6 to 20 $\times 10^{-3}$	1–4 days
Sulphur Dioxide	SO_2	0.03 to 1.2×10^{-3}	1–6 days
Ozone	O_3	0 to 0.05	?

Notes

[a]   This is the atmospheric background concentration, and not the concentrations found in polluted areas. When a range of concentrations is given, it indicates that these have been measured by different workers at different places.

[b]   Minimum concentration of CO_2, measured away from centres of population. In population centres, CO_2 concentrations vary from about 0.034 to 0.035%.

[c]   For photosynthesis. Turnover time with the deep ocean is in the order of centuries.

?   Indicates that little is known about the residence time of the gas.

namely, sulphur dioxide (SO_2), nitrogen oxides (NO and NO_2), carbon monoxide (CO), and non-methane hydrocarbons. Other reactive gases can also cause pollution problems at elevated concentrations, for example, the halogen gases chlorine and fluorine and their acid derivatives hydrochloric and hydrofluoric acid, but these problems tend to be local rather than universal, and their background concentrations an order of magnitude or more less than the gases reported in Table 1.1.

The background concentrations of the reactive gases have remained, to the best of our knowledge, constant with time. This means that the sources and sinks (as the formation and removal processes are generally called) are in balance, and also, for gases with a high pollutant contribution, that the sinks are able to cope with the additional burden from man. This is partly explained by reactivity and partly by the fact that these gases' natural source greatly exceeds the pollutant one in the majority of cases (Table 1.2). It can be seen from Table 1.2 that nature produces over 10 times as much hydrogen sulphide (H_2S), at least an equivalent amount of nitrogen oxides (NO and NO_2, written as NO_x), and over 100 times as much ammonia (NH_3) as is produced by man. Sulphur dioxide (SO_2) looks like an exception to the rule, however. Hydrogen sulphide (H_2S) is ultimately converted to sulphur dioxide (SO_2) in the atmosphere and is, therefore, a source of SO_2. When this is taken into account, together with the different relative molecular masses of H_2S and SO_2, it can be demonstrated that the natural and man-made sources are equivalent.

Table 1.2 Sources of air pollutants

GAS	SOURCE		QUANTITY ($\times 10^6$ tons per annum)	
	Major Pollutant Sources (Anthropogenic Sources)	Natural Sources	Pollution	Natural
• Sulphur Dioxide (SO_2)	Combustion of coal and oil, roasting of sulphide ores	Volcanoes	146	6–12
Hydrogen Sulphide (H_2S)	Chemical processes, sewage treatment	Volcanoes, biological action in swamps	3	30–100
• Carbon Monoxide (CO)	Combustion, principally motor car exhausts	Forest fires terpene reactions	300	> 3000
Nitrogen Oxides (NO_x)	Combustion	Bacterial action in soils	50†	60–270†
Ammonia (NH_3)	Waste treatment	Biological decay	4	100–200
• Nitrous Oxide (N_2O)	Indirectly from use of nitrogenous fertilizers	Biological action in soil	>17	100–450
Hydrocarbons	Combustion, exhausts, chemical processes	Biological processes	88	CH_4: 300–1600 Terpenes: 200
• Carbon Dioxide (CO_2)	Combustion	Biological decay, ocean release	1.5×10^4	15×10^4

†Expressed as tons NO_2

Pollution problems associated with the gases of Table 1.2 (CO_2 excepted) arise not as a result of the magnitude of the man-made (anthropogenic) emission but because this emission is concentrated in the areas where people live and work, and more specifically in the cities of the industrial world. Further, most of the world's industry is concentrated in the northern hemisphere (over 90%), and the great majority of this between the latitudes of 30°N and 60°N (Fig. 1.1). In this region, the anthropogenic emission is considerably more important than the natural one, and the potential effects are more serious because of the concentration of the receptor of most vital interest to us – man himself. Natural emissions are only rarely concentrated in limited areas in this way and when they are, as in the case of volcanoes, their intermittent nature and scattered location minimize the effects of their reactive gas emissions.

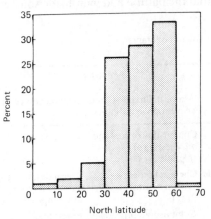

Fig. 1.1 Distribution of energy consumption with latitude for the Northern hemisphere.

With the exception of volcanoes, natural emissions do not fluctuate significantly from year to year, although they may fluctuate considerably within a year. Man's emissions, on the other hand, are steadily increasing as populations and industry expand. The estimated world-wide production of the past century of one such pollutant – sulphur dioxide – is shown in Fig. 1.2. In 1860, this was about 5 million tonnes, mainly from coal combustion. Today, about 190 million tonnes are produced, of which about 100 million is from coal (which generally contains about 0.5 to 4% sulphur), 50 million tonnes from the refining and burning of oil, and the remainder chiefly from the smelting of copper, lead and zinc ores. Similar curves can be drawn for all other anthropogenic gases, despite the increased use of pollution control measures.

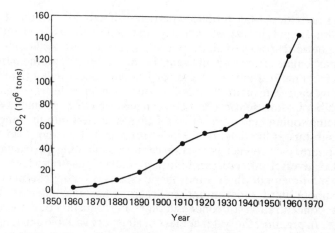

Fig. 1.2 Estimated sulphur dioxide emissions for past years. (From Strauss, W. (ed.) *Air Pollution Control – Part II*, Wiley Interscience (1972).)

In the long term our most serious pollution problems may arise not from the reactive gases but from unreactive emissions, like carbon dioxide, which have no harmful interaction with living systems. Like the reactive pollutant gases, the emission of carbon dioxide is increasing from year to year; this increase, in contrast with the reactive gases, is accompanied by an increase in concentration of the pollutant. Carbon dioxide plays a vital role in the earth's radiation balance and, therefore, as we shall see in Chapter 3, in the earth's climate. The change in the CO_2 concentration, it is predicted, will produce an associated change in the climate of the earth.

Any inert pollutant will increase in concentration with time if there are no sinks available. However, this is rarely the case in practice. For CO_2 the problem is the lack of balance between the sources and sinks over a timescale that is relevant to the atmospheric concentration (Table 1.2). Achievement of equilibrium with the oceans would require centuries if all anthropogenic emission ceased. For that other group of inert pollutants, the chlorofluoromethanes, whose concentration is also increasing, it is the nature of the final removal process in the stratosphere which is the cause of concern. Although the chlorofluoromethanes appear to have no tropospheric sink, they are decomposed in the stratosphere, giving products which react with ozone. Because this removal process exists, the concentration of chlorofluoromethanes at a steady state of usage would eventually reach an equilibrium concentration in the troposphere. However, this would take decades rather than years.

The real atmosphere also contains particulates and vapours, both of which have natural cycles in the atmosphere which are affected by human activities. Man produces particles directly as a result of his agricultural and industrial activity, and indirectly as a result of atmospheric reactions of anthropogenic gas emissions. Although the total particulates from human agents is only about 10% of the particulates from natural sources, these particles are concentrated (like the pollutant gases which are their major source) in the industrial regions of high population density. Particles can have an effect on amenity by producing soiling, and on health by acting as carriers into the lung of trace substances like lead or the polyaromatic hydrocarbons. In the atmosphere, particles act as reaction centres, facilitating gas reactions and also adversely affecting visibility and sunlight penetration.

The major vapour phase atmospheric species is water. This persists as a vapour until supersaturation occurs, when it condenses, the vapour concentration for condensation being a function of the temperature. In practice, the water content of air is very variable and ranges from 0 to about 4% by weight. Many industrial activities emit water vapour directly – thermal power generation for example – and others affect the water cycle indirectly by release of heat to the atmosphere or to large bodies of water. Other substances also exist in the vapour phase in the atmosphere, in particular the non-methane hydrocarbons from both pollutant and natural sources, and sulphuric acid, the product of sulphur dioxide oxidation. Both vapours are associated with the incidence of smog in 'favourable' circumstances.

For gases and particles to be air pollutants we have already said that they must exceed their normal background concentrations to a significant extent. To put this in another way, substances in the air are pollutant when their concentrations are sufficient to have adverse effects on man and his environment. These effects will be dealt with in greater detail in Chapter 3, and for the present they can be summarized as either being direct or indirect. Indirect effects arise as a result of changes in the physical properties of the earth's atmosphere system – the infra red radiation balance in the case of CO_2, and the amount of incoming ultra violet radiation in the case of ozone depletion caused by chlorofluoromethanes (Chapter 3). The potential effects are global in their implication and as such require international rather than national cooperation for their study and, if necessary, for their solution. Direct effects occur as a result of the interaction between a pollutant and a receptor; such effects result from the higher than background concentrations pertaining near the source of pollutants. Direct effects can be immediate if the concentration is high enough (acute effects), or can develop over the long term as a result of continuous exposure to higher than background levels (chronic effects). The extent of area affected distant from the source or sources is a function of the distance the pollutant is transported before the concentration is reduced, by reaction or removal at the earth's surface, to a level at which no direct effects occur.

To date, it is the direct form of air pollution of which we have been most aware, particularly in its most acute manifestation – the London smogs of the 1950s. It was direct pollution, by hydrochloric acid gas from the alkali industry, which led to the first Alkali Act in the UK in 1863 and most control legislation since that date.

Where the pollutant emission is typical of the process, as in the case of the alkali industry, and there is only one plant carrying out the process in the area, then an association between effects and pollutant is readily made and control measures can be taken where necessary. Other examples of unique pollutants from particular processes would be: hydrogen fluoride (HF) from fertilizer plants and the aluminium industry, and odorous reduced sulphides (hydrogen sulphide and mercaptans) from the wood pulping industry (Chapter 2).

The products of fuel combustion such as particulates, sulphur dioxide, nitrogen oxides, carbon monoxide and hydrocarbons are common to most industries and, in addition, have domestic and vehicular sources, making it more difficult to lay blame for observed effects individually. In cities, pollution comes from a multiplicity of sources, not least of which is the motor car, which contributes more than 50% of the hydrocarbons and nitrogen oxides to most urban air-sheds and 90% of the carbon monoxide. These pollutants affect the city producing them most seriously under low wind speeds and stable atmospheric conditions, which prevent the dispersal of pollutants and maximize their opportunities for reaction. Reactants and products form a complex mixture which can have an additive effect on a receptor or result in a significant enhancement of effect over and above what would be expected from the sum of the individual pollutants present (Chapter 3).

Dispersal of Air Pollution and the Weather

Most air pollutants are contained in gases which, being produced by combustion, are much warmer than the surrounding air. As a result, they are buoyant and tend to rise, hot gases being less dense and therefore lighter than cold. As they rise they mix with the surrounding air, becoming progressively colder, and so rise more slowly. The rising of the mixture of air and waste gases containing the pollutants then depends on changes in air temperature with increasing height.

Normally, in the troposphere the air cools as it rises. At high altitudes, on top of very high mountains, it is always less than 0°C and snow and ice are permanent. Even in the tropics there are perpetually snow-capped peaks. If a parcel of air passes from a low level to a much higher altitude and does not exchange heat with the air surrounding the parcel, it expands as the pressure decreases and cools off. This temperature drop is known as the 'adiabatic lapse rate', which for dry air is 1°C per 100 m rise (Fig. 1.3(a)). Conversely, of course, if the parcel of air is compressed as it falls its temperature will rise at a similar rate.

While the theoretical drop in temperature with height is the adiabatic lapse rate, the actual changes can be very different due to the effect of winds, sunshine and topography. If the rate of temperature decrease is greater than the adiabatic lapse rate as in Fig. 1.3(b), a body of warm air such as a smoke plume will rise quickly and the situation is considered 'unstable'. Under these conditions, air pollu-- tants will be rapidly dispersed. However, if the temperature decrease is less than the adiabatic rate as in Fig. 1.3(c), a warm body of air will rise much more slowly, and after some dilution and cooling by expansion will stabilize. Such conditions are considered 'stable' and air pollutants are not dispersed.

With unstable conditions, the warm air (which may contain air pollutants) will rise and be replaced by cooler, clean air from the cold upper layers, giving good mixing. The height to which the warm air rises and mixes with the cooler air until it meets its equal in temperature is called the mixing depth, and is the upper level for air pollution dispersion.

The mixing depth is a function of the seasons, the ground temperature, sunshine and other meteorological factors. On a clear summer day with sunshine, the mixing depth may be several thousand metres, while in winter, with lower ground temperatures and less sunshine, it may be only one or two hundred metres.

It is also possible to get warmer layers of air covering colder ones. Most frequently this can occur near a large body of water such as a big lake or the ocean. When the sun goes down in the evening and the land cools off, a light breeze can bring cool air off the lake or ocean to the land. This forms a cold layer under the warm one, which rises to cover the cold layer (see Fig. 1.3(d)). This is called an 'inversion' layer, and will trap the body of air below it. In valleys between mountains, with steep sides, inversions occur as the morning or afternoon sun warms the upper layers of the air, while at the bottom the air is sheltered and remains cold, or is cooled by a river flowing through the valley.

With low wind speeds the inversion can be quite stable, and partly mixed and cooled air pollutants will not rise into the warm layer. This results in an accumulation of air pollutants below the inversion. In some conditions, for example those that can exist in the Thames Valley through London, inversions can persist for several days and a critical accumulation of air pollutants can form. In Los Angeles, inversions at certain times of the year can occur almost daily, forming at night and persisting into the morning. Inversion conditions are, in fact, quite common and stable inversions occur frequently in the areas of many

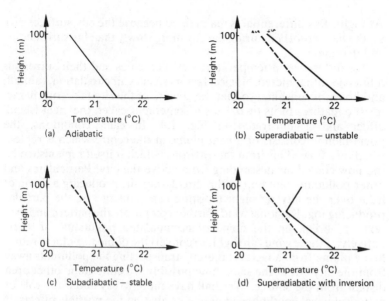

Fig. 1.3   Atmospheric stability as determined by the environmental lapse rate.

major cities which, for historical reasons, are sited next to the ocean or in river valleys where there is an adequate water supply and access for trade by sea.

Cities at high altitudes such as Johannesburg and Mexico City, which lie almost 2 km above sea level, also suffer from air pollution. At these altitudes, the buoyancy of the warm exhaust gases is less, relative to the surrounding air, and ground level inversions are frequent. In cold climates in Northern Europe and in the Northern United States and Canada, much more fuel is burned during winter than summer; winter air pollution levels of some air pollutants, notably sulphur dioxide, are significantly higher than those during summer. In fact, the temperature variations are closely followed by fuel consumption, and in turn by pollution levels.

Cities are 'heat islands' in comparison with the surrounding countryside during both summer and winter. This is most marked in winter in colder climates as considerably more fuel is consumed for heating. The heat is lost from the buildings and passed into the city atmosphere and, it has been found that some European cities average as much as 5°C above the surrounding countryside. During summer cities are also warmer than the surrounding countryside and so, to a lesser extent, are their outlying suburban areas. Summer heat differentials arise from the higher heat absorption of the built environment – roads and buildings – adding to the normal industrial and domestic heat output.

At night, this differential is maintained because the city surfaces give up the heat absorbed during the day more slowly than the surrounding countryside.

The difference in temperature between cities and their surrounds influences other meteorological factors such as air circulation, rainfall, and hours of sunshine, so that the climate of the city takes on characteristic features of its own. This is sometimes called the 'heat island' effect and is illustrated in Fig. 1.4. In calm conditions, the phenomenon consists of a heat plume at the centre which is replenished by a flow of air from the environs, which is itself replenished by the now cooled air descending from above the city. Particulates and other pollutants build up in this circulating air, producing a dome of haze over the city; at night moisture can condense on the particles producing fog. Regional winds further complicate the pattern and may serve to break up the dome of accumulated pollutants, or alternatively, if these winds are light, to convert the dome to a plume rather like a plume from a factory, thereby transporting the pollutants away from the city but at the same time possibly affecting some other area distant from the source. We shall have more to say about the role of meteorological conditions in aiding or abating the possible effects of pollutants in Chapter 6.

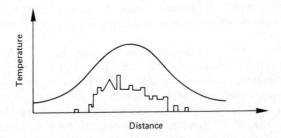

Fig. 1.4   Heat island effect showing the temperature variation over a city.

When pollutants are held over a city for an extended period, either as a result of an inversion or as a result of the city's own local meteorology, the condition known as 'smog' can develop. There are two distinct forms of smog: that typically associated with London, which gave the phenomenon its name (smoke plus fog), and the more recent type associated with Los Angeles, which is entirely different in origin and is generally distinguished as photochemical smog.

The incidence of the classical London type smog has been associated with winter conditions, and arises from the interaction between particles and sulphur dioxide in the atmosphere when the humidity is high, whereas the most severe photochemical smogs have been associated with summer conditions and arise from the interactions between nitrogen oxides and hydrocarbons in the presence of sunlight. We will look at the chemical details of both processes and their effects in Chapter 3.

2 Sources of Air Pollution

In order to control air pollution it is necessary to understand what the sources of air pollution are, and how they operate. It is possible, at least in theory, to control air pollution by eliminating the sources. This, however, would have a most disruptive effect on our society and the way we live. For example, we would not be able to use very much electricity, drive automobiles, or use anything which contains metals or plastics. In fact, life in a contemporary urban environment would be virtually impossible. We therefore have to control the air pollution produced by our activities, which requires a knowledge of the processes which support our life style.

Some sources of air pollution are very large and concentrated: they are the large factories, chemical plants, oil refineries, metal smelting and recovery works, and electric power stations. However, these contribute only one third of the total mass of our air pollution burden (calculated for all pollutants other than CO_2). Transportation in the developed countries contributes about 45%, with space heating, particularly in winter in the colder countries of Europe and Northern USA and Canada, also making a significant contribution. Incineration of refuse adds another 5% to the total burden. Within cities it is the multiplicity of small sources, particularly our private motor cars, that are the main cause of the degradation of air quality.

The major groups of industrial sources are oil refining, metal extracting, and the manufacture of chemicals. These and some other industrial air pollution sources will be considered separately from simple combustion systems used for steam raising for industry and electrical power generation, and internal combustion engines used for transportation.

Petroleum Refining

Our major fuels today, world-wide, are based on petroleum. Crude petroleum is a mixture of liquid hydrocarbons containing, as impurities, from 1 to 4.5% or more sulphur, depending on the source, and a number of inorganic metal compounds. It is generally found in reservoirs in the rock structure several thousand metres beneath the surface of the land, and increasingly beneath the relatively shallow waters of the continental shelves. It comes up through pipes inserted into holes drilled into the subterranean rock structure in order to release the oil and can then be piped, or transferred by large tankers, to a 'refinery' where the crude petroleum undergoes a series of physical and chemical processes which provide different fuels and also feedstocks for the petrochemical industry (Fig. 2.1). This, in turn, produces a huge range of products: synthetic rubber and carbon black; synthetic fibres such as Nylon* and Terylene†; fertilizers; plastics; pharmaceuticals, and many others.

The initial process in the refinery is the distillation of the crude petroleum into a number of fractions (see the 'flow diagram'). Some of these fractions can be used directly, while others have to be given further treatment. The proportion of the fractions obtained in the initial stages depends on the source and nature of the crude petroleum, but the relative proportions of the final products can be changed by using different chemical processes.

The single most important product of a refinery is the comparatively volatile ('light') fraction used for motor spirit (petrol, gasoline). Some of the less volatile ('heavier') fractions can be converted to motor spirit by catalytically cracking the molecules. Another important conversion process is reforming, a process whereby fractions from the distillation column unsuitable for gasoline are rearranged to give more suitable products at high temperature and pressure over a platinum catalyst.

Much of the sulphur in the crude petroleum stays with the heavier fractions. The least volatile of these is asphalt, which is used as a road sealant. Next in volatility are the fuel oils or 'residual oils', followed by the heavy gas oil used as a feedstock in the petrochemical industry, automotive distillate (fuel for Diesel trucks and cars, and also domestic heating oil), 'white spirit' (dry cleaning fluid), and kerosene (jet fuel). The 'lighter' fractions, including motor spirit, are all treated to remove the last traces of sulphur compound impurities as these could lead to rapid corrosion in the engines and other systems in which they are used.

Some residual oils are produced as a result of the operation of the refinery, and must be used or disposed of in some way. Partly they are burned in the steam raising boilers within the refinery, while the rest are sold for combustion in large industrial boilers, for electricity generation, and for ships (bunker oil). There are some processes for removing the sulphur from these residual oils (hydrodesulphurization or 'HDS' processes), but they are expensive and at present are little used except in Japan. Light petroleum crudes with high motor spirit and other volatile fraction components and comparatively little sulphur, such as are found in Libya, Nigeria, Indonesia and Australia, are a premium grade with a much higher market value than the heavy

*Trade name for du Pont fibre   †Trade name for ICI

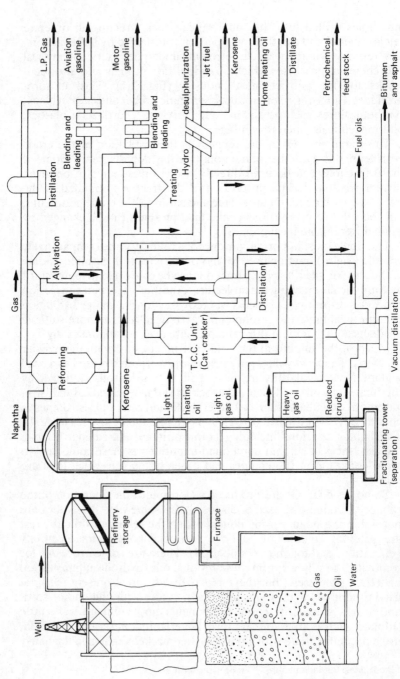

Fig. 2.1   Diagram of a refinery showing typical products.

high sulphur crudes.

The atmospheric emissions from oil refineries are of four types. First, there are hydrocarbon vapours from some of the refining units which are not fully sealed, from leaky valves, from storage tanks, and a multiplicity of other sources. Second are the combustion waste gases from the heaters, boilers, furnaces and flares used in the refinery; these contain sulphur dioxide from the sulphur in their fuel. Third, there are the sulphur containing gases, principally hydrogen sulphide and sulphur dioxide, from units in the refinery which remove sulphur compounds from the distillate products. In many refineries, these gases are used in a special plant (Claus process, Fig. 2.2), which can produce either sulphur or, if required, sulphuric acid. Finally, there are fine particles which come from the catalytic cracker catalyst recovery furnace, which can be the largest source of particulate air pollution in a refinery.

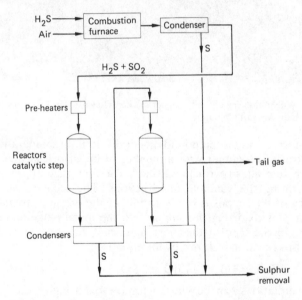

Fig. 2.2 Diagram of a Claus plant for converting refinery sulphide waste to sulphur, which can be used in the manufacture of sulphuric acid.

Smelting of Non-Ferrous Ores

The most concentrated source of sulphur dioxide pollution is from the roasting of non-ferrous sulphide metal ores, the most common being copper, lead and zinc. Significant air pollution damage to vegetation was first observed some 80 years ago around the large copper smelters in places such as Ducktown, Tennessee, and Queenstown in Tasmania (Fig. 2.3).

Fig. 2.3 Vegetation loss around a copper smelter, Queenstown, Tasmania. (Courtesy G. Pratt, University of Melbourne.)

Non-ferrous metal compounds generally occur in comparatively low concentrations together with a number of impurities such as arsenic and iron sulphides (pyrite) and silica. The first stage in processing is concentrating the valuable ore fractions. The second stage in the recovery of the metals is the 'roasting' of the ore in a stream of air, which means oxidizing the sulphide to the metal oxide and sulphur dioxide. In the case of the most significant zinc ore, 'Zinc blende' (ZnS), this reaction is shown by the equation:

$$2\,ZnS + 3\,O_2 = 2\,ZnO + 2\,SO_2$$

Similar equations can be written for the oxidizing of the lead and copper sulphide ores.

For lead, the subsequent treatment is the smelting of the oxide (called calcine) with carbon, in a blast furnace:

$$2\,PbO + C = 2\,Pb + CO_2$$

The crude reduced lead contains dissolved copper, gold, silver, zinc, antimony and other metals as impurities, and has to be refined.

In the case of copper, where the principal ore is chalcopyrite ($CuFeS_2$), the first stage is also roasting the ore with air, which oxidizes some of the copper and produces sulphur dioxide in concentrations of about 8% (Fig. 2.4). The molten products of roasting still contain mixtures of the sulphides and oxides of copper and iron, as well as other non-volatile impurities. These are treated in a reverberatory

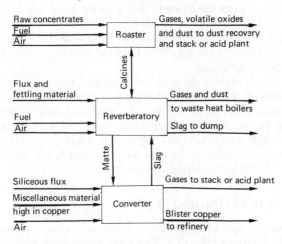

Fig. 2.4 Flow chart for copper smelting processes. (From Stern, A. C. (ed.) *Air Pollution Vol IV*, Academic Press (1976).)

furnace, together with limestone and silica, which combine with the iron and other impurities to form a slag which floats on top of the molten copper and iron sulphide, which remains. The melt is drawn off, and forms, on cooling, the copper 'matte'. Some sulphur dioxide (about 1 to 2%) is given off during the heating process in the reverberatory furnace. In a subsequent process, in a converter, the matte is heated in air and the iron and sulphur are oxidized. The reactions are very complex but essentially the iron oxide is removed as a slag after fluxing with silica, and the copper oxide and sulphide react to give metallic copper and more sulphur dioxide:

$$Cu_2S + O_2 = 2\,Cu + SO_2$$
$$2\,CuO + Cu_2S = 4\,Cu + SO_2$$

The sulphur dioxide concentrations from the converter can reach, intermittently, about 5%.

A possible use for the waste sulphur dioxide from non-ferrous smelting processes might be conversion to sulphuric acid. However, there are two problems with this, even when the sulphur dioxide concentrations are above the required 4% level for the conversion. Many of the processes discussed above are intermittent, while a sulphuric acid plant requires a steady flow of gas. Furthermore, the gases are nearly always contaminated with impurities from the original ore, which act as catalyst poisons in the catalytic converter of the acid plant.

Arsenic trioxide, (As_2O_3), and hydrogen fluoride (HF), are two 'poisons' which are produced in roasting, while in the case of lead and zinc their own oxides act as catalyst poisons. Nonetheless, in many places the gases from the roasters and other processes are converted to sulphuric acid, or in some cases liquid sulphur. Sulphuric acid in some cases is combined with ammonia, and so provides ammonium sulphate for use as fertilizer; or it may be combined with phosphate rock into superphosphate fertilizer.

There are, however, a number of places in which non-ferrous sulphide ores are roasted where there is no accessible market for any sulphuric acid or fertilizer that may be produced, one example being the Mount Isa smelter in Queensland. Here, the sulphur dioxide is emitted through a chimney over 200 m high, which even under adverse meteorological conditions results in a ground level concentration sufficiently dilute to be considered harmless.

Manufacture of Iron and Steel

Iron, unlike the non-ferrous metals, is found in abundance as almost pure oxide, haematite (Fe_2O_3). This is treated directly with carbon, in the form of coke, in a blast furnace (Fig. 2.5). The reaction here is essentially:

$$Fe_2O_3 + 3\,C \rightarrow 2\,Fe + 3\,CO$$

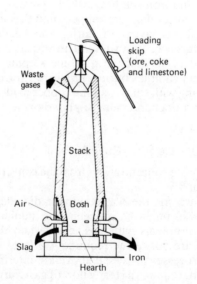

Fig. 2.5 Iron blast furnace.

The carbon monoxide gas is recovered and after cleaning is used as a fuel in other sections of the steel works, for example, in the 'reheating' furnaces where steel billets are heated before rolling into steel plate, wire or other sections. The blast of hot air in the blast furnace is needed to speed up the complex processes of oxidation and reduction.

The iron resulting from this process (pig iron) contains considerable amounts of dissolved carbon (about 4%) as well as other impurities, such as sulphur, silicon, manganese and phosphorus. The pig iron has to be refined to contain precise small amounts of the impurities, which give it the properties of a particular type of steel. This is now generally carried out in large steel converters, in which oxygen is lanced on to the surface to oxidize the unwanted impurities. The resultant steel contains less than 1% carbon and only trace amounts of the other elements. During the oxygen addition, vast quantities of fumes of very fine brown iron oxide are produced, which are potential emissions and require collection.

The process of making the coke used in the blast furnace also produces air pollution. In coking, coal is heated in narrow shaft-like containers, called 'retorts', to 700–1000°C. The coal swells during the coking process and gives off tars and combustible gases. The gases (mixtures of hydrogen, methane and carbon monoxide) are used, in part, to heat the retorts, and in part for other purposes such as 'town gas'. If the filling of the retorts with coal is not carried out carefully, large quantities of dust can be produced. When the coke is ready, it is pushed out on to trucks and quenched with water. This produces large quantities of steam, which also contain particles and some tarry odorous gases. Also during the carbonization, quantities of gases such as hydrogen sulphide (H_2S), and other organic sulphides (mercaptans) which have very objectionable smells, can leak from the coke ovens to the atmosphere particularly if there are poor seals on doors and hoppers. A badly operated, leaky coke oven 'battery' is a serious source of air pollution.

Iron foundries
These are usually relatively small factories, frequently found in the older suburbs and built-up areas, where pig iron and scrap metal is melted down and cast into moulds with very intricate shapes, from engine blocks to cast-iron decorations for houses. Their air pollution problems, dust and fume from materials handling and smelting, are similar to those of steel works but on a much reduced scale. Foundries frequently also have an odour problem resulting from the partial burning of the resins used in holding together the sand moulds. These sand shapes, particularly for the 'cores', use mixtures of pitch, linseed oil and kerosene as a binder, and are baked in a 'coke oven', producing an acrid odour.

Chemical Works and Other Industrial Processes

The factories which make the enormous range of materials, from agricultural fertilizers and pigments for paints and fabrics to synthetic fibres, rubber and plastics, can be classed together as 'chemical' works. These have been, traditionally, one of the major concentrated sources of air pollution when uncontrolled. This is because gaseous reaction products, which were often poisonous or otherwise unpleasant, were released to the atmosphere without regard either to their effect on the environment or to their intrinsic economic value as a source of raw material for other chemicals. These were the first processes to be controlled, initially in England (through the Alkali Act of 1863) and subsequently elsewhere. Chemical works are probably the places where air pollution controls are most strictly enforced. As there are so many commercial chemical processes, only some examples can be considered here.

Sulphuric Acid Manufacture

Sulphuric acid is made (Fig. 2.6) either from pure sulphur (which may have been obtained from natural gas), or from a metal sulphide,

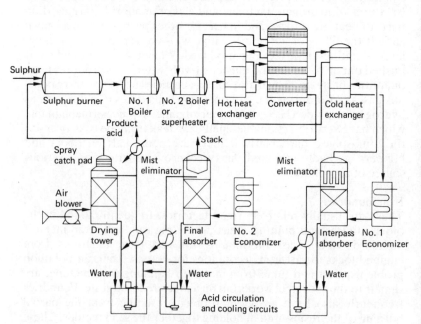

Fig. 2.6   Flow chart for sulphuric acid manufacture. (From Phillips, A., in Thompson, R. (ed.) *The Modern Inorganic Chemicals Industry*, The Chemical Society, London (1977).)

typically pyrites (iron sulphide, FeS_2). When pyrites are roasted in air, this gives:

$$4 FeS_2 + 11 O_2 \rightarrow 2 Fe_2O_3 + 8 SO_2$$

When sulphur is burned, the relation is simply:

$$S + O_2 \rightarrow SO_2$$

The sulphur dioxide from these oxidation processes, generally in concentrations between 8 and 14% in air, is passed over a catalyst at 450°C. The further oxidation

$$2 SO_2 + O_2 \rightleftharpoons 2 SO_3$$

then takes place, the extent depending on the catalyst type and the number of contact passes. The sulphur trioxide is then absorbed in moderately concentrated sulphuric acid to give acid of the required commercial concentration:

$$SO_3 + H_2O \rightarrow H_2SO_4$$

The oxidation of sulphur dioxide is not complete in one pass over the catalyst, and in conventional multi-pass plants (four passes), about 96 to 98% conversion is achieved, the remaining sulphur dioxide being emitted to the atmosphere. In some very modern, large plants with additional conversion stages and partial absorption between oxidation stages, 99.6% or more of the sulphur dioxide is converted to the trioxide and subsequently the acid. For acid plants making some 1500 tonnes each day, the daily emission of sulphur dioxide is therefore reduced from about 40 to 4 tonnes, and 54 tonnes of additional acid is produced.

Fertilizer Manufacture
In the manufacture of 'superphosphate' fertilizer, sulphuric acid is mixed with crushed phosphate rock in enclosed mixers called 'dens'. Some of the constituents in the phosphate rock react and are emitted, the most notable air pollutants from this process being hydrogen fluoride and fluorsilicic acid.

Petrochemicals
There are thousands of different processes used in the petrochemical industry (the most diverse and important branch of the chemical industry), but in virtually all of them a hydrocarbon, or mixture of hydrocarbons, reacts under the influence of heat and pressure and the presence of a catalyst to give a hydrocarbon product. This requires the handling of volatile organic compounds which, under some conditions, escape to the atmosphere. Petrochemical plants use as their raw materials some of the products of oil refineries, and are usually associated with them.

Non-Metallic Minerals

Materials such as limestone for cement, clay for ceramics, asbestos, sand for glass and concrete, crushed rock (aggregate) for concrete and road surfacing, and coal, are mined and quarried in very large quantities. Many of the raw materials have a low unit value and so have to be produced relatively close to the point of consumption; for example, sand and aggregate. With the exception of underground coal mining, these materials are quarried. In the case of coal the quarrying process is called 'open cut' or 'open cast' mining. If the material is hard, it is first blasted, collected and crushed, and then transported to storage bins from which it is carried in trucks to the points of use. Soft coals (brown coal or lignite) are simply cut from the seams. Other harder coals may receive additional treatment, separating out some non-coal constituents by washing.

Cement and glass are obtained by high temperature treatment of limestone and sand respectively, as principal raw materials. For cement manufacture, the crushed limestone, after blending, is calcined in rotary kilns with emission of considerable amounts of fine dust, mainly lime. In glass manufacturing, the sand, soda ash and crushed broken (used) glass (called 'cullet') is melted down in a furnace with the evolution of fumes and gases. In addition to silica, these contain carbonates, nitrates, chlorides and fluorides, and present a specific source of air pollution.

In all these industries the mining, crushing and transport of the materials is associated with the generation of large amounts of dust, both coarse and relatively fine. This contains silica and, in some instances, asbestos, and so is a danger to the workers exposed to it. In these industries, considerable 'in plant' efforts are made to contain the dusts, to maintain 'industrial hygiene'. Careless plant operation and transport with inadequate controls can, however, cause external dust problems.

Paper Pulp Manufacturing

Paper consists of layers of cellulose fibres pressed together, the fibres being obtained from wood. Wood, cut from the forests, is first debarked and chipped. The chips are then treated chemically to free the fibres. One of the most common processes is the 'Kraft' process (Fig. 2.7), where the wood chips are cooked in large steel vessels at comparatively high pressure with an alkaline solution of sodium hydroxide (NaOH) and sodium sulphide (Na_2S). The sodium hydroxide and sulphide interact with the resinous material (lignin) which holds the wood fibres together, and a 'black liquor' is formed containing small quantities of hydrogen sulphide and organic sulphides, which have a very pungent odour even in low concentrations. In a process much used for hardwoods, the 'neutral sulphide semi-chemical' (NSSC)

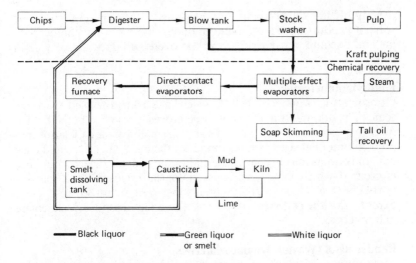

Fig. 2.7 Flow chart for the kraft process for paper manufacture. (From Stern, A. C. (ed.) *Air Pollution Vol. IV*, Academic Press (1976).)

process, the reagent is a less alkaline solution of sodium sulphide (Na_2SO_3) and sodium carbonate (Na_2CO_3), and a similar pungent smelling black liquor is produced.

To recover the chemicals for reuse, the black liquor is dried and then burned in special furnaces, which produces an ash of sodium sulphide and sodium carbonate. Sodium hydroxide is recovered by reacting these two products with calcium hydroxide (burnt lime). The vapours given off during the drying, and some of the volatile compounds incompletely burnt during combustion, contain the sulphide (H_2S), methyl mercaptan (CH_3HS), dimethyl mercaptan ($(CH_3)_2S$), and dimethyl disulphide ($(CH_3)_2S_2$) responsible for the offensive odour associated with paper pulp mills.

Paint Drying
A widespread source of air pollution from industrial and domestic sources is the evaporation of organic solvents during the drying of paint. Not only do some people find the smell of paint objectionable, hydrocarbon solvents participate in photochemical smog formation and contribute to photochemical air pollution. This has led to the substitution of less reactive solvents in paint formulations used in some regions and, where possible, to water based paints.

'Baking' of paint or enamel in motor car body works, electrical appliance factories, can manufacturing, and other similar plants produces a distinctive odour. Paint is sprayed on to metal surfaces and the product is then baked in special ovens where the paint is dried by driving off the solvent, and then hardened. Not only do the organic solvent vapours have a pronounced odour, but they may interact with air and become partly oxidized in the ovens, producing a very acrid smell.

Food Manufacturing
Our society uses many foodstuffs which have been processed on a large scale. The odours from many of these processes, such as bread baking or coffee roasting, when carried out on a small scale are pleasant, but on a commercial scale, where tonnes of the material are treated each day, the odours may become penetrating and pungent. Thus, a nuisance is created which may have a psychological effect. Coffee roasting particularly produces a sharp smell which nauseates. So, to a lesser extent, do the odours from bread baking and even chocolate manufacture.

Rendering of (Waste) Organic Materials
Rendering or cooking of waste materials in pressure vessels also produces unpleasant odours. Rendering is carried out on natural materials of high protein content, either specially harvested or produced for the process or obtained as waste products from some other industry, such as meat works. The processing drives off ammonia (NH_3) and organic amines (methyl amine, ethyl amine, etc.) and also fats. The high protein product is used as poultry feed. The edible fats which are a by product in meat rendering are used in the food industry, e.g., biscuit making, while inedible fats are used for making soap. It is the ammonia and amines plus a small quantity of organic sulphides which give the nauseating odour to the vapours from rendering.

Incineration of Refuse
Incineration, or burning of wastes, is a frequent cause of air pollution if not controlled. Examples which are seen in many of our cities are the burning of old tyres, of insulation around used electricity cables (in order to recover the copper), of leaves and garden cuttings, and of household wastes in small domestic incinerators. When these are carried out without proper combustion equipment, considerable amounts of smoke, and frequently obnoxious odours, result. In many cities, private incinerators and open burning are now prohibited.

When garbage is disposed of by combustion in large, well-designed incinerators with adequate controls, there are no air pollution problems. 'Landfill', i.e., burying the refuse in old quarries or in trenches in open fields is a widely used alternative. It is essential that the garbage is covered daily with soil to avoid odours and vermin. The filled land can later be used for other purposes such as parks or playing fields. If buildings or other substantial structures are to be placed on the filled land, special precautions have to be taken with foundations. Another difficulty is that rain water can penetrate the soil and leach soluble constituents out of the buried garbage unless the ground at the site is impermeable to liquid penetration. These chemicals can find their way into the ground water or nearby rivers, leading to long term water pollution.

An alternative to incineration and landfill is the recycling of the wastes; thus, old rubber tyres can be 'reclaimed' and turned into shoe soles or similar products, or they can be ground and used as road filler. Paper can be converted into cardboard or reprocessed to other forms of paper. Metal and glass can to a large extent be recycled, while domestic 'putrescible' wastes can be composted and turned into organic fertilizer.

External Combustion Systems

Combustion of fossil fuels (coal, petroleum and natural gas) is the most common source of air pollution from industry. Fuels are either burnt directly on site to produce heat or steam, or burnt elsewhere, generally near the fuel supply, to produce the electricity used by industry. Fossil fuels are also used directly for domestic cooking and heating, or indirectly for the same purpose in the form of electricity.

Until comparatively recently, the major fuel used by man was wood. It was not until the late 19th century in the USA and Europe that fossil fuels equalled wood in importance, and for many parts of the world even today wood is still the predominant fuel. Emissions from wood include particulates, carbon monoxide, hydrocarbons, ammonia, nitrogen oxides and, of course, carbon dioxide. Although none of these are sufficiently large to affect the worldwide anthropogenic totals, they can be very important locally, for example, in Christchurch, New Zealand where particulates from open fires for household heating are a significant contribution to winter fogs.

The greater heat value per unit mass of coal resulted in its replacing wood for domestic heating and cooking, and for the direct firing of boilers in the industrial sphere. Coal was, in addition, the first transport fuel of importance in the age of steam, and the first source of gas for fuel. Although oil and natural gas have supplanted coal for many of these uses, coal is still the major fuel in electricity generation. This use of coal is unlikely to decline in importance as the alternatives of using oil or gas for this purpose, or escalating nuclear power programmes, are becoming increasingly unacceptable. Fortunately, the known reserves of coal are very large so that, even in pessimistic terms, we will have adequate supplies for several hundred years. However, coal has certain disadvantages that make it less attractive than oil and gas. In particular, its energy content per unit mass is not as high, its mining causes damage to the landscape, and it is dirty to handle and to burn, giving off large amounts of non-combustible particulate as ash and other pollutants. The disadvantages are minimized for electricity generation by using the coal on site and transporting the electricity, the site always being some distance removed from urban conglomerations.

There are many types of coal, ranging from what are called 'low rank' coals, with a high proportion of volatile matter (such as peat, brown coal or lignite), through bituminous coals with less volatile content, to the anthracite coals which have a very low volatile and moisture content and a high proportion of carbon. The heating value of coals follows the rank and is perhaps twice as high for high rank bituminous coals and anthracite as it is for lignite. In general, high rank coals are found at much greater depths and are mostly mined in underground mines, while 'brown' coals (lignites) are generally much more accessible and are mined by open cut methods.

Of course, coal is not just a compound of carbon, hydrogen and oxygen; it also contains small amounts of sulphur and nitrogen, as well as traces of mineral impurities, which together are responsible for many of the air pollution problems in coal combustion.

When coal is burned with air, a series of complex reactions take place, which can be summarized by the following scheme.

The carbon combines with oxygen to give the oxides of carbon:

$$C + O_2 \rightarrow CO_2$$
$$2C + O_2 \rightarrow 2CO$$

With the right conditions, the carbon monoxide (CO) will oxidize to the dioxide (CO_2):

$$2CO + O_2 \rightarrow 2CO_2$$

The combined hydrogen in the coal oxidizes to water:

$$2H_2 + O_2 \rightarrow 2H_2O$$

At high temperatures, which occur in the flame, both nitrogen in the coal and nitrogen in the combustion air react with oxygen to give nitric oxide (NO):

$$N_2 + O_2 \rightarrow 2NO$$

(The NO subsequently combines with oxygen in the atmosphere to give NO_2).

The sulphur in the coal oxidizes to the oxides of sulphur, mainly sulphur dioxide:

$$S + O_2 \rightarrow SO_2$$

and about 1 to 3% of the sulphur dioxide formed is further oxidized to the trioxide under the excess air conditions required to burn coal efficiently:

$$2\,SO_2 + O_2 \rightarrow 2\,SO_3$$

The mineral impurities in the fuel are incombustible and form ash. Part of this, 'fly ash', becomes airborne and is carried out of the boiler in the waste gases. In older boilers, lumps of coal were burned on a grate and much of the ash fell through the grate (or off the end of the grate if it moved) and into the ash pit. So in these, with good firing practice, only a small amount of the ash escaped to the air. In large modern boilers, however, the coal is pulverized and burned as a fine powder, which results in much of the ash being carried with the waste gases into the atmosphere unless an effective cleaning system has been installed.

The main purpose of very large scale coal combustion today is for generating electricity. Coal is injected in pulverized form into the boiler combustion chamber, where it is burnt at high temperatures. The flame radiates heat to the boiler tubes surrounding the walls of the furnace, where the water is heated and evaporated. The hot combustion gases pass through banks of tubes, the first being the superheater where steam is heated at high pressure to temperatures far above the boiling point. The hot gases are then passed through water preheaters and air heaters before being channelled through a cleaning system and out to the atmosphere through a chimney. Thus, as much heat as possible is extracted from flame radiation and the hot gases.

Electricity is generated as an alternating current by passing the superheated high pressure steam through turbines coupled to alternators. After passing through the turbine, the steam is exhausted to large, water-cooled condensers where it is condensed to water which is pumped back to the boiler.

Modern power station boilers, burning hundreds of tonnes of coal each day, are very large structures resembling externally a tall multi-storey building. These boilers are very effective combustion systems. There is, therefore, virtually no carbon monoxide in the waste gases, and the fly ash contains very little unburnt carbon. Modern methods used to remove the fly ash are, as will be shown in Chapter 5, efficient and little escapes so that even when the boiler is operating at full capacity no smoke is visible. It is the smaller and older boilers which are less efficient in their combustion, and are frequently without any means of fly ash control, which can give rise to serious smoke problems.

In spite of the virtually complete combustion, and the relatively high efficiency (85%) of heat transfer to water and steam, there are unavoidable thermodynamic losses in converting the energy in the steam into mechanical and, in turn, electrical energy. The waste gases from power stations, which are usually at temperatures between 120 and 180°C, contain some of the heat which was contained in the original fuel. Considerable amounts of energy are also lost when the water is condensed, the cooling water used in the condensers giving up heat either in cooling towers or to a river or the ocean. Overall, at least 65% of the energy which was in the fuel is rejected or lost, and cannot be converted into useful electrical energy by the conventional power station.

However, it is possible in some industrial situations, notably in the paper, chemical and oil industries, or with district heating of houses (used in New York, parts of London, Germany, Sweden and elsewhere), to use the residual thermal energy in the steam which is normally rejected to the cooling water. This is done by 'back-pressure' turbine systems. Steam, at the required temperature and pressure, is drawn from the turbine (called a 'pass-out' turbine) and led to the process needing the steam, with the residual steam being passed to the condensers. With this system, about 75 to 80% of the energy in the fuel can be used.

Some of the best examples of energy utilization are in the pulp and paper mills. Here, electrical energy is generated to drive the machinery and large quantities of steam are used to heat the pulp digestors, the black liquor driers, and the paper driers.

Oil fired power stations are very similar to the coal fired units, except that the oil is fired into the boiler in finely atomized sprays, the spray droplets behaving in some ways like the particles of coal in the pulverized coal fired units. The fly ash from oil fired boilers is far less than from coal fired units, and only rarely are the particles collected.

The reactions which occur are similar to those in coal fired units. Large oil fired boilers usually use a heavy fuel oil with a fairly high sulphur content (2 to 4%), so there is an appreciable amount of sulphur dioxide in the waste gases.

Smaller boilers such as are used for industrial or commercial steam and hot water, and domestic heating, are much simpler in construction. Today, they are usually gas or oil fired and in the latter case use lighter oil fractions which have very low sulphur or mineral content. These can be fired so as not to produce smoke or other air impurities, particularly when the burners are well adjusted and correctly sized for the heating.

Gas fired boilers are very efficient combustion units and virtually pollution free, except for small amounts of nitrogen oxides formed during combustion from combustion air nitrogen. In all boilers, good design can reduce the nitrogen oxides formed.

In all the units described, the firing takes place in a chamber which is surrounded by tubes containing the heat transfer medium, which is usually water or steam. However, in the chemical and petroleum industries a high boiling point liquid mixture of diphenyl and diphenyl oxide (in the ratio of 1 to 3), called Dowtherm after the makers Dow Chemical Company, which boils at 250°C at atmospheric pressure, is often used, while the nuclear power industry has been investigating liquid sodium (boiling point 880°C at atmospheric pressure) as a heat transfer medium for new reactors. Because the heat source is separate from the heat transfer medium, the term 'external combustion' is used for these boilers and furnaces, in contrast to the 'internal combustion' systems which will be discussed below. A comparison of the emissions resulting from the use of the three major fuels in electricity production is given in Table 2.1.

Benzo(a)pyrene figures are also given in Table 2.1 as representative of the polyaromatic hydrocarbon group of compounds whose carcinogenic properties make them suspect as possible causes of environmental cancers.

Table 2.1 Comparative emissions from electricity generating plants using different fuels (units are grams per 10 000 BTU's unless otherwise specified)

	NO_x	SO_2[1]	Particulates[2]	Hydrocarbons (as methane)	Benzo(a)pyrene[3] (g per 10^{12} BTU's)
Coal:					
Bituminous	3.36	6.38 S	0.5 to 2.7 A	0.034	18 to 222
Anthracite	3.63	6.89 S		0.036	
Residual Oil	3.15	4.75 S	0.24	0.097	33
Natural Gas	1.77	0.002	0.068	negligible	not available

Carbon Monoxide emissions are low for efficiently fired boilers, < 0.08 g per 1000 BTU's with all fuels.

Heating values used for the fuels given were:
Coal: 2.7×10^7 BTU per tonne bituminous
            2.5×10^7 BTU per tonne anthracite
Residual Oil: 1.5×10^5 BTU per gallon
Natural Gas: 10^3 BTU per ft³

(Source: Bond, R.G. and Straub, C.P., *C.R.C. Handbook of Environmental Control, Vol I Air Pollution* (1974).)

[1] S is the percentage of sulphur in the fuel.
[2] A is the percentage of ash in the fuel. The range of figures is a function of the dependence of the emission on the method of combustion.
[3] Benzo(a)pyrene emission is a function of the boiler sizes and design.

Internal Combustion

Most methods of obtaining a traction force in a moving vehicle (with the exception of steam locomotives, 'steam cars' and electric vehicles) use an internal combustion engine. In such an engine, a mixture of a combustible vapour is mixed with air, compressed in a cylinder with a piston, and then ignited. The energy produced by the combustion drives the piston forward, and this in turn moves a connecting rod which drives a crankshaft. The two common forms of internal combustion differ in the method of igniting the combustible mixture. In one type the mixture is ignited by an electric spark, so this is called the spark ignition or Otto engine, after its inventor (Fig. 2.8). It is commonly used in automobiles and motor cycles. In the other form,

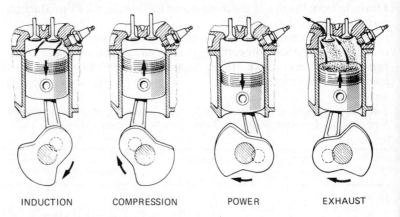

| INDUCTION | COMPRESSION | POWER | EXHAUST |

Fig. 2.8   The operating cycles of the Otto engine.

the mixture is ignited spontaneously by a high compression and the heating which results from this. This is the Diesel engine, also named after its inventor (Fig. 2.9). An understanding of the operation of these engines is necessary if one is to grasp how air pollution from this source can be reduced.

In the conventional spark ignition engine, the fuel is the easily vaporized petroleum distillate product of oil refineries, which is called gasoline, petrol or, in Europe, benzine. This is pumped from the fuel tank through a jet where it is vaporized and mixed with air in the carburettor. The air-fuel mixture is then passed, in turn, to one of a number of cylinders. These are usually arranged either 4, 5 or 6 in line, or 8 or 12 paired off in a V configuration.

In the cylinder (Fig. 2.8), the mixture is compressed until the piston reaches the top position, at which time the spark ignites the mixture which drives the piston outward. The power in the working stroke is transmitted to the crankshaft and flywheel, which stores the energy, uses some for the compression and exhaust strokes in the cylinder and the remainder for driving the wheels of the vehicle. The burnt gases are removed from the cylinder by the return movement of the piston, with the second valve open. In a normal engine the process is repeated about 1500 times each minute or 25 each second.

The Diesel engine (Fig. 2.9) differs from the spark ignition engine in several significant ways. First, only air is compressed and not an air-fuel vapour mixture. The compression is adiabatic, meaning that it is carried out without any of the heat of compression being carried away. Close to the end of the compression the fuel is injected into the air heated by the compression, and ignites spontaneously. The hot gas expands and drives the piston outward, and this is the working stroke. This is followed by the exhaust stoke, during which the exhaust valve is opened. Towards the end of this the new fresh air comes into the cylinder, helping to flush out the residual, burned gases.

In the spark ignition engine, combustion starts from one point only and proceeds unevenly and because the cylinder walls are cooled the combustion is incomplete. Some of the combustion gases pass along the cylinder walls past the piston and enter the crankcase. Typically, of the hydrocarbons emitted, some 60% pass out through the exhaust pipe, 20% are lost through the crankcase, and the remaining 20% are lost equally from the fuel tank and the carburettor. The other air pollutants, chiefly carbon monoxide and the oxides of nitrogen and particulate containing lead and condensable hydrocarbon species such as benzo(a)pyrene, are emitted through the exhaust pipe. The lead compounds come from the 'anti-knock' additives used to make the modern high compression engine run smoothly, reduce corrosion of the valves, and extend engine life.

The quantities of pollutants in the exhaust gases from a particular spark ignition engine are also a function of the size, age and the operating characteristics, the 'tuning' of the engine, and whether the engine is cold or warm. For example, an engine tuned to a 'lean' mixture (containing more air relative to the fuel) produces less carbon monoxide and hydrocarbons in the exhaust than an engine tuned 'rich'. When starting a cold engine, we 'pull out the choke', which reduces the quantity of air admitted and produces a rich mixture.

When the engine is running normally, the emissions also change with the driving mode, hydrocarbon and carbon monoxide emissions being greatest for a decelerating vehicle whereas nitrogen oxide emissions are greatest under accelerating and cruising conditions.

Table 2.2 shows typical average emissions for motor vehicles running on petrol and diesel fuels. As can be seen from this Table, for uncontrolled vehicles the emissions of the major pollutants, carbon monoxide, hydrocarbons and nitrogen oxides, are higher for the Otto engine. The major exception is the higher level of particulate emissions from the diesel engine, although this is a function of the load on the engine and can vary by a factor of 20 from a value less than the Otto engine at three-quarter engine load to a maximum value at full load.

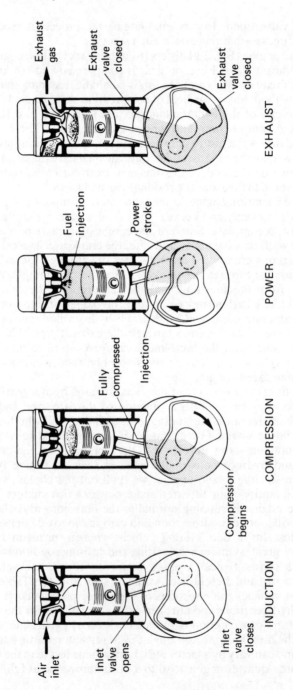

Fig. 2.9   The operating cycles of the Diesel engine.

Table 2.2 Typical average gaseous emissions in automobile exhaust (units given are grams per kilometre)

Pollutant	Otto Engine	Diesel Engine
Carbon Monoxide	60.00[a]	0.69 to 2.57[d]
Hydrocarbons	5.90[a]	0.14 to 2.07[d]
Nitrogen Oxides	2.20[a]	0.68 to 1.02[d]
Particulates	0.22[b]	1.28[b]
Sulphur Dioxide	0.17[b]	0.47[b]
Lead	0.49[b]	–
Benzo(a)pyrene	14×10^{-6}[c]	24×10^{-6}[b]

An average fuel economy of 10.3 km/1 was assumed for light duty diesel engines.

[a] Average values for pre-controlled cars on standard USA Federal driving cycle, US EPA *A Computation of Air Pollution Emission Factors* (1974).
[b] Calculated from figures given in Bond, R.G. and Straub, C.P. *CRC Handbook of Environmental Control, Vol. I, Air Pollution* (1974).
[c] Butler, J.D., *Air Pollution Chemistry,* Academic Press (1979).
[d] US EPA figures – cars tested Nissan, Datsun, Mercedes, 220D, Peugeot 504D and Opel Rekord – reported in Stern, A.C., Ed. *Vol. IV, Air Pollution,* Academic Press (1977).

Sulphur dioxide is also emitted in greater amounts by the diesel engine but has, to date, been considered a minor component of transport emissions.

Automobile emissions contain a greater number and variety of hydrocarbons than are found in the original fuel, including the aldehydes, organic acids and benzo(a)pyrene. Aldehydes and organic acids are oxidation products of combustion and in the case of the aldehydes are particularly active precursors of photochemical smog. The suspected carcinogen benzo (a) pyrene and the family of poly-nuclear hydrocarbons it represents are products of pyrosynthesis during combustion from constituents of the fuel. In the diesel, the polynuclear hydrocarbons are associated with the soot and can be much lower than the figure given for the Otto engine if the timing and air to fuel ratios are correct, and much greater than the average figure given if the engine is improperly maintained.

Aircraft emissions also contain carbon monoxide hydrocarbons, nitrogen oxides and particulates. Of the three types of aeroplane engine in use today, the newer jet and turboprop planes have much lower levels of all the major emissions than the piston engined aircraft. The only exception is the high level of particulate emission associated with conventional jet planes. Although planes make little difference to the total air burden overall, their contribution in the vicinity of major airports, such as Chicago, can be significant. Plane emissions during flight are generally considered to be emitted at a high enough level and be sufficiently dispersed to be unimportant. It is considered possible that supersonic aircraft nitrogen oxide emissions in the stratosphere may have a deleterious effect on the ozone concentration. We shall return to this topic later.

Comparison of pollution sources

When discussing air pollution from different sources, particularly combustion processes, it is important to compare the relative quantities of air pollutants produced. In Table 2.3 the average quantities of the major pollutants produced for 1 tonne of fuel consumed are shown.

Table 2.3   Emissions from combustion sources (kg per tonne fuel)

| | INTERNAL COMBUSTION | | EXTERNAL COMBUSTION | | | |
| | | | Fuel Oil | | Coal | |
	Otto Engines	Diesel Engines	Power Generation	Commercial Domestic	Power Generation	Commercial Domestic
Carbon Monoxide	395	9	0.005	0.025	0.25	25
Nitrogen Oxides	20	33	14	10	10	4
Sulphur Oxides	1.55	6.0	20.8 S	20.8 S	19 S	19 S
Hydrocarbons	34	20	0.42	0.26	0.1	5
Aldehydes and Organic Acids	1.4	6.1	0.08	0.25	0.0025	0.0025
Particulates	2[a]	16[b]	1.3[c]	1–12[c]	8 A[d]	2–8 A[d]

S: To obtain sulphur oxides produced, multiply number by % S in fuel (S = sulphur)
A: To obtain particulates produced, multiply number by % A in fuel (A = ash)
(Note: most coal fired units are fitted with control devices, and emissions are of the order of 1 to 10% of these)

[a]   About 20% of this total is lead from the anti-knock additives.
[b]   Largely unburnt carbon, on which traces of aromatic compounds, such as 3,4 benzopyrene and similar compounds, have been adsorbed.
[c]   Largely metal oxides from impurities in the fuel oil. These include the oxides of iron and vanadium.
[d]   Metal chlorides and sulphates (calcium, magnesium, iron, etc.) from ash inclusions in the coal.

Most obvious is the large amount of carbon monoxide (a function of the combustion method) from the Otto engine compared with the Diesel engine, and the insignificant amounts of this produced by external combustion systems. Nitrogen oxide emissions are also a function of the mode of combustion, and are most easily controlled in external combustion systems. Ash and sulphur dioxide emissions are not, of course, primarily a result of the type of combustion but of the type of fuel used. Motor spirit is a highly refined product containing very little sulphur and no ash. However, it has been seen that the production of refined fuels from crude petroleum inevitably results in some residual fuels, and these also need to be disposed of, usually by combustion in a manner resulting in a minimum of air pollution.

This comparison does not show the importance of either the quantity of fuel burnt which produces the pollutants, or the dispersion which dilutes them and reduces their effect. A very large power station, providing enough electricity for a city of one million people (say 1200 MW), burns about 12 000 tonnes of coal every 24 hours, while a medium-sized car can travel about 7000 km on one tonne of motor spirit. Again, a reasonable assumption is that one motor car uses about 2 tonnes of motor spirit each year. In a city with a population of 1 million in Australia or the United States, there would be 3 cars for every 10 people, or 300 000 vehicles consuming 600 000 tonnes of fuel each year. Unlike the power station, they release their pollutants at ground level and tend to concentrate this during the morning and afternoon peak periods on 300 days each year. This produces a relatively greater impact than the generation of electricity, which could be well away from the population centre, has carefully monitored air pollution controls, and a tall stack to aid dispersion of the residual pollutants.

3 The Effects of Air Pollution

The gaseous and particulate materials added to the atmosphere by the activities of man are considered to be pollutant when their concentrations are sufficient to produce harmful effects. The majority of man-made emissions to the atmosphere also have natural sources and in many cases these are larger than the pollutant ones (Table 1.2). The world as we know it developed in the presence of these chemicals which cannot, therefore, be considered to be harmful per se but only if they produce unacceptable effects at concentrations above the natural background level. The effects that are of concern are those that do, or may in the future, affect man's health and well-being and his enjoyment of the world as we know it without undue alteration of biological or physical systems.

In practice, the association between effects and pollutant concentrations – air pollution criteria – is not clearcut because of the immense number of variables involved. This lack of adequate criteria adds to the problem of decision making on the concentration levels of pollutants acceptable in air, as we shall see in Chapter 6.

After formation, air pollutants are emitted to the atmosphere and dispersed. Once mixed with the air, some air pollutants, such as the inert fluorinated hydrocarbons used in aerosol sprays, persist unaltered and become mixed throughout the atmosphere where they potentially have a global influence. More reactive pollutants have a shorter lifetime in the atmosphere and are removed either by conversion to normal atmospheric constituents or by deposition on the surface of the earth. In the process they may react with other atmospheric constituents to form secondary pollutants, which are removed by the same process. Both the primary emission and the secondary pollutant can cause alteration to the chemical composition of soils and waters, and direct damage to biological systems and property. The mix of pollutants in the air is never a constant or a simple one and the damage observed in a particular situation is often the result of more than one pollutant acting together. In certain cases, a synergistic interaction occurs where the total effect is enhanced over and above the sum of the effects of the individual pollutants present. Because our interest in pollutants is man-centred, effects on human health will predominate in the discussion which follows.

Information about effects can either be obtained where the effects occur, in the field, or in the laboratory under controlled conditions. Both approaches have some merits and neither is completely satisfactory. Field studies – epidemiological studies – require the correlation of pollutant concentrations with other observations, such as changes in the mortality of a population or the frequency of incidence of a particular disease. This is intrinsically very difficult because so many other factors that have a bearing on mortality and disease vary within a population. There is, however, no other satisfactory way of determining the actual risk experienced by all elements of a population, particularly those most at risk – the young, the old and chronically ill.

Laboratory studies can provide information under controlled conditions on the effect of pollutants on materials, plants, animals and, to a limited extent, on humans. The laboratory approach provides for controlled conditions and the removal of unwanted variables, and allows for the unequivocal association of level of effect with concentration. Laboratory methods are not suitable for the determination of elevated concentrations on the health of weak members of society for ethical reasons, nor for practical reasons are they suitable for determining the effects of exposure to long term low level concentrations. These important areas as far as human health is concerned are only accessible through epidemiological studies. Studies in controlled conditions on animals do provide useful information on the physiological mechanisms of pollutant action, although the concentration level at which these occur cannot be extrapolated to the human population. There are two sources of information on human exposure to elevated levels of pollution which could perhaps be classified as unintentional experiments. These are air pollution episodes where, as a result of meteorological conditions, a whole cross section of population is exposed to high concentrations for several days, and industrial accidents, where an area and its population is exposed to a single chemical or simple mixture, such as occurred recently in Seveso, Italy, when there was an accidental release of TCCD*. Information is also available on health effects from occupational exposures. Such exposures often occur over a long period of time, and some of the variables of the group exposed are removed as age, social background and state of health are more uniform in the workplace than among the population at large. Other variables are also reduced, the mixture to which a work force is exposed will be simpler, and length of time and concentration will be better quantified. Such information is, of course, useful as it helps quantify the effect of specific pollutants or pollutant mixes on healthy adults. On the other hand, the group exposed is not

---

*TCDD   2,3,7,8 trichlorophenoxyacetic acid, a by-product of the manufacture of the herbicide 2,4,5-T

representative and the concentration and mix of pollutants is not what the population at large is in contact with. In particular, possible synergistic interactions will not be included and the pollutants of greatest interest in the ambient air are not those for which data is available from the workplace.

The operation of synergism and the complexity of a polluted air system where acute effects have occurred, effects whose association with air pollution is not disputed, are best seen in an examination of air pollution episodes.

Air Pollution Episodes

'Classical' Smog

The most serious air pollution episodes are the classical smogs so closely associated with London. These episodes all occurred in cold winter months in areas where coal was a major fuel for both industry and heating. They are all characterized by high levels of sulphur dioxide and smoke particulate which built up under stagnant weather conditions lasting three days or more. Mortality and illness were significantly increased and, in the worst case in terms of the numbers affected, in London in the week of 5-11 December 1952, there were 3500 to 4000 more deaths than the expected average for that time of the year. Those most susceptible were the elderly and the young, particularly individuals with a history of cardio-respiratory illness.

No single primary pollutant could be singled out for blame in any of these incidents and it is probable that a synergistic association between sulphur dioxide, particulates and their reaction product sulphuric acid was operative. These particulates are often referred to as smoke particulates, or simply smoke, because this relates to the standard method of measuring air particulate burden at that time, i.e., by the degree of darkening of a standard filter through which a known volume of air had been passed. Sulphuric acid in classical smog is the product of the dissolution of sulphur trioxide in water, the sulphur trioxide being formed by the oxidation of sulphur dioxide on particle surfaces. This oxidation has been found, in laboratory studies, to be catalysed by carbon and certain metal ions, such as iron, magnesium and vanadium, all of which would be expected to be present in smoke particulates. Nitrogen oxides are similarly oxidized and hydrolysed to nitric acids. Concentrations of the two primary pollutants in London in 1952 rose to a maximum for sulphur dioxide of 1.46 ppm daily average associated with a daily average for smoke of 4.5 mg m^{-3}. This concentration of SO_2 is considerably less than the level of 5 ppm below which no measurable effect has been observed on healthy adults, and which has been chosen as the threshold limiting value† by most western countries. Synergism is further supported by the effectiveness of the smokeless zone regulations in London in preventing a recurrence of such episodes even though the average annual sulphur dioxide concentration has increased on occasions.

The health effects of long term exposure to sub-acute levels of sulphur dioxide and particulate are better documented than those for any other pollutant or group of pollutants. Numerous studies have demonstrated correlations with one or both these pollutants and respiratory tract disorders in both adults and children. Unfortunately it is not possible to quantify the results because of the different measures of respiratory disorder used and particularly because of poor information about the concentration of pollutants to which subjects were exposed. Generally only one of the pollutants was measured and often residence location was taken as a measure of the level of pollution to which an individual had been exposed. Residence location is an unsatisfactory measure of pollutant exposure because it correlates with other unrelated factors like socio-economic status, and it takes no account of the possibility that much of the subject's time is spent away from his residence.

The studies do point to an association between SO_2 particulate pollution and chronic bronchitis, emphysema, and asthmatic attacks. There is also evidence of an association of these pollutants with the incidence of upper respiratory tract problems in children. The effect is greatly aggravated in adults by cigarette smoking and is possibly synergistic. It should also be added that the relationship between cigarette smoking and respiratory disorders is stronger than the relationship with air pollution.

'Photochemical' Smog

Photochemical smog was first recognized as a problem in Los Angeles in 1943, and has since been detected in many cities of the world. Photochemical smog refers to the complex mixture of products formed from the interaction of sunlight with two major components of automobile exhaust, nitric oxide (NO) and hydrocarbons. Other pollutant species present in the atmosphere such as sulphur dioxide (SO_2) and particulates can also be participants, as can hydrocarbons and nitric oxide from other sources, but they are not essential for the production of the characteristic high oxidant levels associated with photochemical smog formation.

Smog is further favoured by stable meteorological conditions, when the urban emissions are held in the urban airshed by an inversion acting rather like a lid over a reaction vessel, maximizing contact and

---

† The threshold limiting value, TLV, is the limit for 8 hour industrial exposure of healthy workers over a lifetime.

reaction whilst preventing dispersion. The severest incidents occur when this inversion is steady for several days, allowing for further emissions and reactions adding to those of the previous days. The oxidant formed is predominantly ozone with varying amounts of other oxidizing components, including the notorious peroxyacetyl nitrates (PAN). These minor components contribute to a variation in the observed effects for the same total oxidant concentration in different places. The variations in oxidant composition occur as a result of variations in the mix of hydrocarbons initially present and the length of time the particular parcel of air has to react. These factors vary from airshed to airshed and from day to day within the same airshed.

The level of total oxidant concentration achieved, despite the variations in effects, is still the best measure available of the severity of photochemical smog formation. The maximum oxidant concentration achievable for a particular day is a complex function of hydrocarbon and nitrogen oxide concentrations. This relationship is shown for several US cities in Fig. 3.1. Although the concentration values may

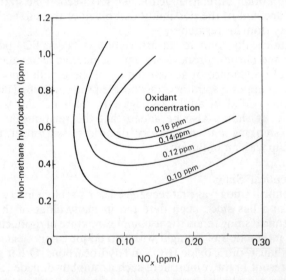

Fig. 3.1 Relationship between nitrogen oxide, non-methane hydrocarbon concentrations and oxidant in Philadelphia, Washington D.C. and Denver during 1966-8. (From US, EPA Document AP 84 (1971).)

vary in other countries, it is suspected that the shape of the relationship will remain the same. It can be seen that a decrease in hydrocarbons – usually characterized as non-methane hydrocarbons because of the lack of reactivity of methane – does not necessarily produce a decrease in the oxidant concentration maximum at constant nitrogen oxide concentrations. Similarly, a decrease in nitrogen oxide does not produce an improvement in oxidant at constant hydrocarbon concentration in all cases. Failure to take account of the complexity of this relationship in Los Angeles led to control measures which initially worsened rather than improved the existing situation. The importance of obtaining an understanding of the relationship operating before undertaking control measures is obvious.

Despite these complexities, the chronology of smog formation is universally very similar and is shown for a typical Los Angeles day in Fig. 3.2. A morning build-up of ingredients to a peak corresponding to rush hour traffic is followed by conversion of NO to nitrogen dioxide (NO_2) and a subsequent build-up of oxidant and other smog products such as PAN and aldehydes. The pattern of Fig. 3.2 has been reproduced in the laboratory by irradiating surrogate mixtures. For simple cases, where the mixtures consisted of air, nitrogen oxides and one or two hydrocarbons only, successful computer simulations of smog formation have been achieved. The agreement obtained between theory, experiment and the real situation indicates that the basic

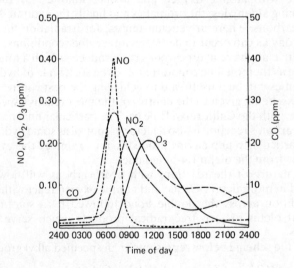

Fig. 3.2 Average one hour concentrations of photochemical smog reactants and products on a typical Los Angeles day. (From US, EPA Document AP 84 (1971).)

reactions are reasonably well understood. However, the complexity of the atmospheric mixture – over 100 hydrocarbons are present in a typical urban airshed – and the lack of information about rates and products, makes computer simulation of a real situation from basic data impossible at present.

The initiating reaction in smog formation is the interaction of sunlight with nitrogen dioxide to give nitric oxide (NO) and atomic oxygen (O).

$$NO_2 + Sunlight \rightarrow NO + O \qquad (1)$$

The atomic oxygen reacts with oxygen (O_2) and a third non-reacting body M, such as nitrogen (N_2) or another oxygen molecule, to give ozone (O_3).

$$O + O_2 + M \rightarrow O_3 \qquad (2)$$

In the absence of any other reactive species, this ozone reacts with NO, completing the cycle with no net production of oxidant.

$$NO + O_3 \rightarrow NO_2 + O_2 \qquad (3)$$

In the presence of hydrocarbons, a whole series of other reactions with atomic oxygen and ozone becomes possible and the above equilibrium is upset.

Olefins, that is hydrocarbons containing a double bond, are most reactive with atomic oxygen and ozone, followed by the oxygen containing aldehydes, the aromatics and finally the saturated aliphatic hydrocarbons. There are circumstances, for irradiation times longer than a day as can occur under stagnant weather conditions, when the unreactive aliphatic hydrocarbons can produce as high a total oxidant reading as their olefinic counterparts. Even so, a scale of hydrocarbon reactivities can be a useful aid to achieving the best control strategy, emphasis being given to the control of the most reactive species, as is the case with the Californian Rule 66. The particular final product of a hydrocarbon's reaction is also a factor in control as some hydrocarbons give particularly unpleasant products, for example, the eye irritant acrolein from the olefin 1,4-butadiene.

The nature of the reactions with hydrocarbons will now be considered in more detail. Ozone and atomic oxygen react with an olefin by addition across the double bond followed by a splitting of the original olefin to give free radical products which serve as chain carriers.

R in the scheme below represents an unspecified alkyl group:

$$RCH = CH_2 +_3 \rightarrow RCHO + CH_2O_2 \qquad (4)$$
$$RCH = CH_2 + O \rightarrow RCH_2 + HCO \qquad (5)$$

RCH_2 and HCO are free radical species, and CH_2O_2 is a di-radical species, all of which are reactive and will react further. RCHO is an aldehyde product which is itself subject to oxidation and further reaction. The olefin shown could have split across the double bond in reaction (4) to give formaldehyde CH_2O_2 and the di-radical $RCHO_2$.

A similar scheme could be postulated for reaction (5), with CH_3 and RCHO as products. Apart from the aldehyde products which have been detected in smog, all the other species are postulated only, their presence being evidenced by the final products observed in the atmosphere and in smog chamber reactions.

The intermediate radicals RCH_2 and CH_3 then react further with atmospheric components:

$$RCH_2 + O_2 \rightarrow RCH_2O_2 \tag{6}$$

$$RCH_2O_2 + NO \rightarrow RCH_2O + NO_2 \tag{7}$$

$$RCH_2O + O_2 \rightarrow RCHO + HO_2 \tag{8}$$

where RCHO is an aldehyde product and HO_2 is a reactive radical species that converts NO to NO_2.

$$HO_2 + NO \rightarrow NO_2 + OH \tag{9}$$

The hydroxyl radical product is an oxidant species of major importance in the photochemical smog process and will be considered separately.

An exactly parallel series of reactions can be postulated for CH_3; the net result in both cases is the formation of aldehyde, the conversion of two molecules of NO to NO_2, and no loss of oxidant. Reactions (7) and (9) explain the conversion of NO to NO_2 which is necessary before any oxidant build-up can occur. This is clearly shown on Fig. 3.2 and is a result of the affinity of NO for O_3 (reaction (3)), which effectively mops up all the ozone as it is formed unless the NO concentration is kept low.

The oxygen containing radicals HCO and RCO are also reactive chain carriers:

$$RCO + O_2 \rightarrow RC(O)O_2 \tag{10}$$

$RC(O)O_2$ represents the peroxyacetyl radical; the oxygen in brackets is double bonded to the carbon as it is in RCO and the aldehydes. $RC(O)O_2$ is the precursor to the eye irritant PAN (peroxylacetyl nitrate), a minor unstable but important product in photochemical smog formation.

$$RC(O)O_2 + NO_2 \rightarrow RC(O)O_2NO_2 \text{ (PAN)} \tag{11}$$

The radical also takes part in other reactions:

$$RC(O)O_2 + NO \rightarrow NO_2 + RC(O)O \tag{12}$$

$$RC(O)O \rightarrow CO_2 + R \tag{13}$$

where R is a free radical. This can react further:

$$R + O_2 \rightarrow RO_2 \tag{14}$$

$$RO_2 + NO \rightarrow NO_2 + RO \tag{15}$$

cf. reactions (6) and (7).

RO reacts in the same way as the oxygen radical product in reaction (8), giving an aldehyde product and the HO_2 radical.

$$RO + O_2 \rightarrow R\,'O + HO_2 \tag{16}$$

The reactions of HCO are somewhat different because of its size and instability. The products, however, are similar to those of reactions (8) and (16).

$$HCO + O_2 \rightarrow HO_2 + CO \tag{17}$$

enabling the chain to continue with conversion of NO to NO_2. Decomposition may also occur but the outcome is still the same.

$$HCO \rightarrow H + CO \tag{18}$$

$$H + O_2 + M \rightarrow HO_2 + M \tag{19}$$

The di-radicals CH_2O_2 and $RCHO_2$ probably undergo rearrangement and decomposition reactions as follows:

$$RCHO_2 \rightarrow RCO + H_2O \tag{20}$$

$$\rightarrow ROH + CO \tag{21}$$

$$\rightarrow RH + CO_2 \tag{22}$$

All three reactions (20) to (22) terminate the chain.

It can be seen from the reactions given that once smog formation processes are underway there are several routes, usually involving peroxides, that keep the NO concentration low and enable the oxidant to build up. These reactions have not proved sufficient to explain the initial rate of conversion of NO to NO_2 and to do so it is necessary to introduce a third oxidizing species, the hydroxyl radical (OH). Hydroxyl can convert NO to NO_2 by a mechanism involving carbon monoxide CO which is in plentiful supply in automobile exhaust.

$$OH + CO \rightarrow CO_2 + H \tag{23}$$

The H reacts according to reaction (19) to give HO_2, and HO_2 converts NO to NO_2 by reaction (9) with no loss of OH. The initiation reactions producing OH, like those for the production of ozone, are photochemical. The two strongest contenders here are the photolysis of formaldehyde and the photolysis of nitrous acid. Formaldehyde is not only a product of smog reaction but is produced as a result of the oxidation of naturally occurring methane, and is therefore always present in the atmosphere.

$$CH_2O + Sunlight \rightarrow HCO + H \tag{24}$$

HCO reacts with oxygen according to reaction (17), giving HO_2, and H reacts with O_2 by reaction (19) to give the same product. The net result via reaction (9) is the conversion of two molecules of NO and the production of two hydroxyl radicals for every molecule of formaldehyde destroyed.

The reaction of nitrous acid, HONO, to produce OH is hypothetical because the presence of HONO in the atmosphere has never been demonstrated, although the equilibrium constant for the reaction

$$NO + NO_2 + H_2O \rightarrow 2HONO \tag{25}$$

has been measured. Nitrous acid decomposes with sunlight to give a hydroxyl radical

$$HONO + Sunlight \rightarrow OH + NO \tag{26}$$

Hydroxyl oxidation of hydrocarbons occurs at a faster rate than the accompanying O_3/O oxidation, either by abstracting hydrogen atoms from the hydrocarbon or by addition when a double bond is present. The abstraction reaction is applicable to aldehydes and paraffins as well as to the olefin example given:

$$OH + RCH = CH_2 \rightarrow RCH = CH + H_2O \tag{27}$$
$$\rightarrow RC = CH + H_2O \tag{28}$$

$RCH = CH$ and $RC = CH$ are free radical products which can then act as chain carriers, adding oxygen in much the same way as the hydrocarbon free radical products of O_3/O oxidation (reaction (6) and following reactions). The addition reaction could proceed as follows:

$$OH + RCH = CH_2 \rightarrow RCH(OH) + CH_2 \tag{29}$$
$$\rightarrow RCH + CH_2OH \tag{30}$$

Postulated reaction sequences involving the addition of oxygen follow in the same pattern as for the other hydrocarbon free radical products.

Smog formation and oxidant build-up eventually stop as a result of loss of sunlight initiation at nightfall and dispersion of reactants and products. The chain reactions cease, partly as a result of dispersion and partly as the result of the removal of the important chain carriers through the following termination reactions:

$$HO + HO_2 \rightarrow H_2O + O_2 \tag{31}$$
$$2HO_2 \rightarrow H_2O_2 + O_2 \tag{32}$$
$$HO_2 + NO_2 \rightarrow HONO + O_2 \tag{33}$$
$$HO + NO + M \rightarrow HONO + M \tag{34}$$
$$HO + NO_2 + M \rightarrow HONO_2 + M \tag{35}$$

HONO is, of course, unreactive in the absence of sunlight. The series of reactions is not intended to be exhaustive but to represent the pattern of reactions that occur and their inter-relationships.

The effects of photochemical smog at the levels so far experienced are probably most serious in relation to plant life, and this will be dealt with in a subsequent section. At present, no unequivocal evidence has been obtained linking photochemical smog incidence with human mortality. Los Angeles studies have shown no difference between excess death rates on photochemical smog days and days without smog but with similar temperatures. There are, however, physiological changes which do occur during smog episodes that can be linked to the high level of oxidant and oxidant products. These effects include eye irritation, sore throats, and impaired respiratory function. Because of the highly oxidizing nature of the smog atmosphere, any sulphur dioxide present is converted into sulphuric acid. This both contributes to the aerosol content of the atmosphere and acts in concert, if not synergistically, with the oxidant in producing the observed human reactions. As previously indicated the onset of effects occurs at different total oxidant levels in different locations. The onset of eye irritation, for example, is observed in Rotterdam at levels that cause no similar effect in Sydney. Such differences have been attributed to minor constituents of greater potency, possibly also acting synergistically. There is some evidence for Los Angeles and Tokyo that the photochemical smog causes an aggravation of existing chronic respiratory conditions but this is not definitive.

One clearly observable photochemical smog effect is reduced visibility. This can be due to solid particles, (e.g. smoke), and aerosols such as sulphuric acid. A cloud of particles and droplets absorb light, which reduces visibility to some extent, but more importantly scatters light, i.e. deflecting the direction in which light travels. This effect is due to the particle cloud between the observer and the distant object, which scatters the light from the sun and other parts of the sky and reduces the contrast which the observer perceives. The duration of this depends on how long the particle cloud remains before it can be removed by wind, or, if the particles are sufficiently large, until they can settle out. When particles are above 15 micrometres diameter the rate of settling is sufficient to remove a cloud in several hours, but a major problem is caused by the smaller particles which do not settle out in a reasonable time. These particles are also the most difficult to remove from the waste gas stream of industrial processes.

Although the reduction of visibility depends on aerosol particulate size and the size distribution, it has been shown experimentally that the visibility (in km), V, varies inversely with the concentration, M (micrograms of aerosol per cubic metre), according to:

$$V = K/M$$

The value of the constant, K, is found to vary experimentally between 900 and 3600 km g m^{-3} (average value 1800 km g m^{-3}). So, on the average, with a particle burden of 100 g m^{-3}, the visibility will be about 18 km, but with three times this burden, which is not uncommon in polluted cities, the visibility is reduced to less than 6 km.

Humidity can also be correlated with visibility because it affects the production of aerosols. For relative humidities greater than 70% it has been found in Los Angeles that there is a rapid reduction in visibility from 6 km to 3 km (at 80% relative humidity). At 90% relative humidity, the visibility is further reduced to 2 km. However, the exact effect of humidity on visibility is a function of the air pollutants present, which will differ for each city.

Effects of Individual Pollutants on Man

Man, living in the developed and as a consequence polluted communities, normally presents only limited areas of skin to the atmosphere but each day he inhales about 7500 litres of air so that his lungs and respiratory system are in contact with, and have the potential to retain, whatever harmful substances might be contained in that air. We have already seen in our considerations of air pollutant incidents that it is the nose, throat and bronchial system which are most often affected. This is true of other pollutants as well; even if the lung is not specifically the target organ it provides the route whereby the pollutant enters the system.

Air enters the nose, where fine hairs filter out most particles greater than about 10 micrometres diameter. The air is warmed and humidified, and is then passed through the windpipe into the bronchial tubes, which subdivide the air stream and pass it into the lungs where there are a multiplicity of air sacks (alveoli). It is in this section of the lung that oxygen (and air pollutants) can be absorbed and transferred to the blood stream (Fig. 3.3).

Readily soluble air pollutants, such as gaseous sulphur dioxide, can be absorbed on the moist walls of the upper respiratory system, but fine particles and droplets – in the range from 0.1 to 5 micrometres diameter – together with any gases adsorbed on these, can be drawn right into the lung and deposited on the lung surface. The dangers of some small particles such as silica and asbestos, which are common in mines, quarries and some industrial plants, are well known. They lead to specific occupational diseases such as silicosis or asbestosis – silicosis being referred to as 'miner's lung' – and great care needs to be taken to protect workers in these industries. Although the techniques of such care are established, various factors, e.g., economic or psychological, can lead to their being neglected. Similar precautions, often very stringent and restrictive to operatives, are needed in industries where heavy metals, beryllium, uranium and other substances are handled.

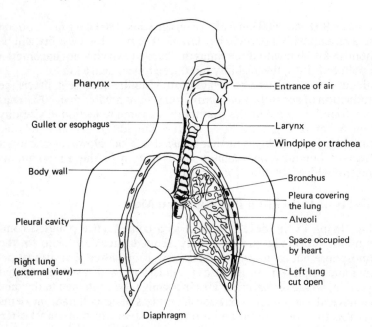

Fig. 3.3   The human respiratory system.

Limits are set as to the maximum concentrations to which healthy persons are permitted to be exposed for periods of up to 8 hours per day. These are called Threshold Limit Values (TLV), although in older literature they are referred to as Maximum Allowable Concentrations (MAC), and are specified for almost all chemicals, minerals and dusts. In the atmosphere, such high concentrations are seldom met, but on the other hand the exposure of a population is over 24 hours rather than 8 hours and all members of that population, including the weakest, are involved. The effect of exposure to long term, low level concentrations of specific pollutants will now be considered for those pollutants, not already dealt with, which are universally present in city air.

Nitrogen Oxides

There is no unequivocal epidemiological evidence that nitrogen dioxide has any adverse effects at the concentrations commonly encountered in polluted air. A most comprehensive study on school children in Chattanooga, Tennessee, purporting to show a relationship between NO_2 concentrations and respiratory disease has since been convincingly questioned. There is, however, some evidence from laboratory tests on asthmatics that exposure to concentrations of NO_2 of between 0.11 and 0.2 ppm for one hour produces changes in airways' resistance in these subjects although the asthmatics were not aware of the changes that had occurred.

Indirect evidence of the way that nitrogen oxides might affect man is available from animal studies, which demonstrated a higher susceptibility to bacterially transmitted disease in exposed animals. This effect has not been demonstrated in man.

Carbon Monoxide

Carbon monoxide is the one pollutant which produces a change in human physiology that can be directly related to the concentration to which the subject was exposed. Blood carboxyhaemoglobin (COHb) can be predicted from atmospheric CO concentrations when account is taken of the level of activity of the subject and the height above sea level. This makes the quantification of concentration and effect much easier.

Death occurs in humans exposed to concentrations of around 1000 ppm, corresponding to blood levels of 60% COHb. Impaired function occurs at much lower blood levels, between 10 and 20% COHb, and has been reported by some workers for levels as low as 2% COHb. Actual levels for urban dwellers in moderate to highly polluted cities vary from 0.8 to 3.7% COHb for non-smokers, to 1.2 to 9% for smokers. Smokers are, therefore, more likely to suffer impairment of function than non-smokers. Although this may reduce the ability of the individual to get the most enjoyment out of life, it does not necessarily shorten his life span.

On the other hand, reasonable correlations have been found between daily mortality levels and CO. In addition, heart function has been shown to be altered by elevated COHb, as evidenced by the electrocardiograms of exposed healthy adults. Because CO blocks the transport of oxygen in the bloodstream, and those with certain heart diseases require a high oxygen supply, it is not unreasonable to draw the same parallel as has been made with cigarette smoking and heart disease, and suggest that CO is a contributing factor.

Heavy Metals

Lead is the most prevalent heavy metal pollutant in the atmosphere and, apart from a few lead-associated industries of local importance, this lead comes from autombile exhaust. Automobile originated lead is associated exclusively with particles in the respirable range, that is, with particles predominantly 1 μm in diameter or less which can most readily reach the lower portion of the lung, the alveoli, where the lead is available for exchange into the bloodstream. It has been shown experimentally that at least 20% of inhaled lead can be so absorbed.

Inhaled lead is an additional contribution to the total body burden of lead which predominantly comes from ingestion. For urban dwellers the inhaled source could be a significant added burden, particularly for those whose exposure is high as a result of occupation, or residence location. In such circumstances, the inhaled source, it has been estimated, could contribute 30% or more to the total.

In contrast to these estimates, a 1968 WHO study of average blood levels in different countries found no clearcut association between high levels and urbanization, New Yorkers having similar levels to residents of New Guinea and somewhat lower levels than found in Finland.

The acute effects of lead poisoning on humans include irritability, motor nerve paralysis, anaemia, miscarriage, and in children defects of the nervous system including mental retardation, cerebral palsy and atrophy of the optic nerve. Acute effects are generally observed as a result of ingestion. Of greater interest as far as air pollution studies are concerned, is the effect of long term exposure to sub-acute levels. This is an area fraught with controversy but there is increasing evidence that children with high lead levels have restricted mental development and a higher incidence of behavioural disorders. The effects are attributed to the irreversible inhibition of the growth of the nervous system of young children by lead. It remains to be established what the source of the high body burden of lead in some children is, but it can be argued that the consequences are of sufficient seriousness to warrant erring on the side of caution. By far the easiest way to cause an overall reduction in lead intake is to desist from adding it to the air as opposed to removing it from food and water.

The other heavy metal universally present in the atmosphere is cadmium. The acute effects of cadmium inhalation include kidney and lung damage, and the so-called itai-itai disease (literally ouch-ouch) characterized by severe pains in the joints and multiple fractures as a result of bone thinning.

There is evidence that long term exposure to sub-acute levels of cadmium leads to a higher incidence of cardiovascular disease and hypertension. Normal atmospheric levels of cadmium which, like lead, is predominantly ingested, are not sufficient to cause any concern, but future increases should be carefully monitored because of the long half-life cadmium has in the body. In this regard, cigarette smokers are at significantly greater risk, inhaling 0.1–0.2 μg of cadmium per cigarette.

Carcinogenic Compounds
One of the main difficulties in establishing cause and effect relationships for cancer is the time delay, usually 20 to 40 years, between the exposure and onset of the disease.

The possibility of an association between air pollution and lung cancer was first realized in 1936 and since then numerous studies have been carried out to establish the existence of an urban factor in lung cancer. The attribution of such an urban factor to air pollution is complicated by the association between cancer and smoking. The association is further complicated by such factors as changing life expectancy, greater exposure of city populations to transmittable diseases, stress, genetic background, socio-economic status, and occupation, all of which could contribute to the differences observed in the incidence of urban and rural lung cancer.

The evidence that air pollution is a causative factor is based on the levels of known carcinogens, in particular the polyaromatic hydrocarbons, found in urban air. These hydrocarbons are strongly linked with skin and scrotal cancer in several industrial situations such as wax processing, high temperature distillation of coal, and in the not so distant past, chimney sweeping. Extracts of atmospheric particulates containing these compounds have been shown to produce similar cancers when applied to the skin of laboratory animals, confirming the carcinogenic properties of the compounds. The link between inhalation and the development of cancer is made by drawing a parallel with cigarette smoking, where the same family of hydrocarbons are implicated.

It has not, however, been possible, over and above demonstrating that there is an urban factor, to associate this factor with the concentration of suspect pollutants. The urban factor does not correlate with the concentrations of polyaromatic hydrocarbons, either as a group or separately. It is also puzzling to find that, although female lung cancer rates are associated with smoking, it has not proved possible to demonstrate an urban factor for non-smoking women. The case for an association between air pollution and lung cancer must, therefore, rest as unproven with a strong indication that more information is needed on the range of suspected carcinogens in the atmosphere and their biological effects, both individually and in combination with other carcinogens and other non-carcinogenic pollutants with which they could act synergistically.

Radioactivity
Future development of nuclear power raises the question of the prospective exposure of humans to ionizing radiation additional to what they already receive from natural and man-made sources. Ionizing radiation is a known hazard, causing direct tissue damage at high exposure levels and inducing cancer and genetic damage at lower long term exposures. It is not known whether there is a threshold level of exposure below which no effects occur.

There are three types of ionizing emission that are of major interest in air pollution studies. These are: alpha(α)radiation, beta(β) radiation, and gamma (γ) and X rays. Alpha radiation consists of energetic helium nuclei (^{4}He), that is, positively charged particles consisting of 2 neutrons and 2 protons. Beta radiation consists also of charged particles, either negatively charged electrons or positrons with the opposite charge. Gamma and X rays are both electromagnetic radiations similar to light but of much higher energy, gamma rays having energy in the MeV range with X rays typically in the 10 to 100 keV range.

The amount of damage that can be caused by ionizing radiation is a function of the amount of energy of the radiation, the target material it is absorbed into, and the nature of the radiation itself. The original unit of dose, the roentgen, is the quantity of X or gamma (γ) radiation that produces one statcoulomb of charge in 1 cm^3 of dry air at STP. Converted to energy units, 1 roentgen is equal to 87 ergs per gram of air. The value of a roentgen for other materials does not equal 87 ergs per gram, and consequently a new unit, the rad, proved necessary. The rad is a unit of absorbed dose, where one rad is equal to 100 ergs per gram of absorber. This is still not a satisfactory measure to be related to biological effects because of the variation in the interaction of ionizing radiations with biological materials. For example, heavy ionizing particles like the products of atomic fission cause damage within a short path length and are consequently more dangerous when absorbed than a gamma ray, which transfers its energy over a longer path length. The absorbed dose, in rads, is therefore multiplied by an appropriate factor to take account of the biological effectiveness. The new unit is known as the rem, the roentgen equivalent man where

$$1 \text{ rem} = 1 \text{ rad} \times Q$$

Q, the quality factor, is a function of the energy deposited per unit length of track in the medium. Gamma and X rays have a Q value of 1, beta radiation has values of 1 to 1.7, and for alpha particles Q equals 10. Of these three types of radiation, alpha particles have the lowest penetrating power and therefore external sources of alpha radiation rarely penetrate the body. On the other hand, an internal source of alpha radiation, such as might arise from ingested or inhaled material, is potentially the most dangerous because all the radiation damage is concentrated in the small volume surrounding the alpha emitter. Of the external sources, gamma rays are potentially the most harmful because of their high penetrating power, followed by the lower energy X rays. To account for the distribution of radiation, internal and external, to the body, and the fact that exposure of certain body areas, such as the extremities, is less critical than exposure of other areas, such as the internal organs or the gonads, the dose is further modified to give the dose equivalent in rems,

$$DE = D \times Q \times N$$

where DE is the dose equivalent in rems, D is the dose in rads, Q is the quality factor, and N the product of all the other modifying factors including the distribution of the exposure.

Acute effects of radiation exposure from short term whole body irradiation result principally from the destruction of body cells. Although the effect varies from individual to individual, typically a dose of 25 rems to the whole body results in slight temporary changes in white blood cell counts, and doses of 200 rems lead to nausea, fatigue and possible death. The 50% (LD50) figure for mortality in an exposed population occurs at around 400 rem.

Acute effects are the immediate result of accidental exposure to high level concentrations of ionizing radiation; of greater interest in air pollution studies are the chronic or long term effects on subjects exposed to sub-acute levels over a period of time. These effects of ionizing radiation are conventionally divided into two categories: (1) somatic effects resulting directly from the dose received, and (2) genetic effects, which are passed on to future generations. As with all attempts to correlate dose with effect for low level exposures of human populations, there is a dearth of information, most of the experimental work being done on animals. The largest group of irradiated human subjects that have been studied are the survivors of the Hiroshima and Nagasaki bombings of 1945. Further information comes from studies of the people of the Marshall Islands, who were accidentally exposed to bomb debris in 1954, and from those who have been occupationally and medically exposed before the adverse effects of such exposure were fully realized.

Some connections are well established, for example, the association between leukaemia and ionizing radiation, although the body dose necessary to produce leukaemia is not satisfactorily quantified. This difficulty in quantifying the association of a particular cancer with dose holds for other cancers whose causative connection with radiation exposure is also generally accepted. Lung cancer, to take a further example, is associated with external radiation exposure of the victims of Hiroshima, and also with internal sources of radiation among uranium miners, without the dose effect relationship being established. Overall, this lack of hard data constitutes a continuing problem in establishing safety margins for human exposure to radiation sources.

Ionizing radiation can cause both gene mutations and chromosome abnormalities which result in a higher incidence of abnormal offspring, possibly carrying the genetic defect to yet another generation. Genetic effects require much longer periods of study than somatic effects for their elucidation, and can probably result from lower levels of exposure.

What constitutes a safe level of exposure for a human population? If one accepts that there is a linear relationship between damage and

Table 3.1   Estimated annual average whole body dose rates in the United States

Source	Dose Rate (mrem per year)
Natural background:	
Cosmic rays (varies with altitude)	45
Rocks, soil and buildings (varies with location)	44
Radionuclides in the body (mainly ^{40}K)	18
TOTAL NATURAL SOURCES:	107
Man-made or induced radiation:	
Medical and dental X rays	72
Radiopharmaceuticals	1
Radioactive fallout	4
Occupational exposure	0.8
Miscellaneous (television receivers, air travel, etc.)	3
Nuclear power plant	0.003 to 0.01
TOTAL MAN-MADE SOURCES:	81
TOTAL:	188

(Source: Hodges, L., *Environmental Pollution*, Holt, Rinehart and Winston (1973).)

exposure – that is, that there is not a threshold below which no damage occurs – then every increase in exposure of a total population will produce a statistical increase in the incidence of cancer. This contention is difficult to prove, even with animals, because of the low levels of exposure involved and the length of time required for the experiment, most cancers taking of the order of one-third of a lifetime to develop. Any calculations based on extrapolating known data to lower concentration levels could, therefore, reasonably be assumed to be an upper limit. Such a calculation has been carried out by the US National Academy of Sciences, who estimated that a total exposure of 5000 mrems, corresponding to 170 mrem per year over 30 years, would increase cancer death rates in the USA by 1 to 5%. The most recent US standard (1971) for exposure of individuals outside new power plant boundaries allows for an additional exposure not exceeding 5 mrem per year, which allows for a considerable safety margin on the Academy's estimates. In the worst possible case of an individual living in the direction of the prevailing wind, even on an old power plant boundary, the estimated exposure level is less than this. The average exposure for the population at large is much less again, and fades into insignificance when considered alongside the average exposure from all ionizing sources, both natural and man-made (Table 3.1). As we shall see in Chapter 7, the real concern with nuclear power plants is not their normal day-to-day operation but the possibility of abnormal or accidental occurrences in the plant or associated with the fuel production and waste treatment stages.

Odours
Odours are in a different category to the poisonous gases and particles because they are unpleasant rather than toxic. Odours can be detected by our sense of smell in very small concentrations, lower than those which are amenable to present methods of analysis. While odours can sometimes cause nausea, it is often possible for those exposed to them to build up a resistance to particular odours, so that they can no longer detect them. This frequently happens to workers in the industry producing the smell, where they are exposed to quite high odour levels. There are usually more public complaints about odours than any other form of air pollution and they remain one of the most difficult forms to eliminate because of the low levels to which they have to be reduced to be imperceptible to sensitive people.

Effects on Vegetation

The acute toxic effects of some air pollutants, particularly sulphur dioxide, on plants in the vicinity of non-ferrous smelters roasting sulphide ores have been observed and recognized for a long time. In some instances the destruction of vegetation has been followed by soil erosion, which has prevented recovery. In fact, the widespread damage by hydrochloric acid gas released during the manufacture of sodium hydroxide by the LeBlanc process ('saltcake') from salt led to the Alkali Act in England in 1863, the first systematic attempt to control industrial air pollution.

Air pollution can affect plants to varying degrees. At the lowest levels, i.e. below the 'threshold' there is no effect, such as visible damage, cumulative chronic effects, genetic effects or even gradual changes in the composition of the plant community. However, even at this level air pollutants can be stored in the plants and introduced into the food chain, affecting animals which eat the plants.

Plants take up air pollutants either directly, through interchange of gases with the atmosphere, or through the moisture taken up from the soil. The soil may have been exposed to the air pollutants that were then dissolved in water in the soil. Acidic air pollutants in particular are easily dissolved in surface moisture or rain. Even when the air pollution source has been removed the dissolved materials can continue to affect the plants grown there, although in time they are diluted and are leached out by rain.

More direct is the entry of gaseous pollutants into the plants through

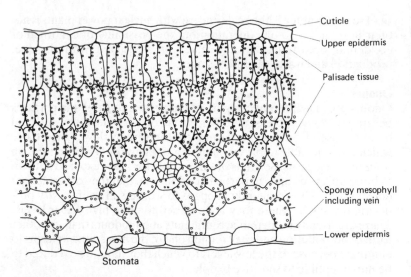

Cuticle

Upper epidermis

Palisade tissue

Spongy mesophyll including vein

Lower epidermis

Stomata

Fig. 3.4 Diagram of a typical leaf cross section.

open stomata (see Fig. 3.4) on the underside of the plants. These are active in the exchange of gases – oxygen, water vapour and carbon dioxide – with the surrounding atmosphere. Gaseous air pollutants, after entering the plant tissue, dissolve in the intercellular water. The resulting acid (if it is an acidic pollutant) then attacks the cell structure within the leaf. This is why easily soluble air pollutants – sulphur dioxide, hydrochloric and hydrofluoric acids, etc. – are the most toxic.

Solid particles are less toxic to the plant, as they are deposited on the hard, waxy upper surface of the leaves. They must then dissolve in moisture deposited on the surface and enter the plant interior through the stomata or through damaged sections on the leaf surface. Some solid pollutants can 'dissolve' in the wax on the surface to enter the plant. Thus solid air pollutants tend to localize their effect and do not harm the whole leaf structure to the same extent as gaseous pollutants. Of course solid pollutants can enter the food chain if their carriers are consumed by animals.

If plants of decorative value have sensitive leaves, then particulate deposits can seriously damage their aesthetic or commercial value, as is the case with flowers, although they may not affect the ability of the plant to survive and reproduce.

The sensitivity of plants, and animals for that matter, varies with the type of pollutant and its concentration, but also depends on whether more than one pollutant is present; two or more pollutants can reinforce each other (synergism) and increase or decrease the effect of the individual pollutant.

The length of time of exposure is a further factor. The product of the time – in hours – and the concentration – in parts per million, or other appropriate units – is sometimes called the 'exposure factor'. With most pollutants this varies with the exposure of the plant to sunlight or humidity, because these can affect the opening of the stomata, and the gas interchange that takes place.

When carrying out plant exposure experiments to test sensitivity to a particular pollutant, it is desirable to work under the most sensitive conditions for the plant. Not only do different plants react differently to the different pollutants or mixtures of pollutants, but different species (types) within a genus (grouping) show varying sensitivity.

Information about relative sensitivities is useful in two ways. Less sensitive species can be grown in regions likely to be affected by air pollutants, such as roadside plantings or mid-city parks. Here, resistant varieties can be planted which will grow well under difficult conditions. On the other hand, very sensitive species can be grown near new air pollution sources, or those which have installed control measures. A study of these plants can indicate whether the control measures taken have been effective. Some effects of specific air pollutants are given in the following sections.

Sulphur Dioxide

This gas, on entering the intercellular tissue, reacts with water to give sulphite ions ($SO_3^{2=}$). These are 30 times as destructive as sulphate ions ($SO_4^{2=}$). Initial symptoms are a darkening of the affected parts of the leaf, followed by flaccidity (indicating internal breakdown of the cell structure) and a drying out to white, dead tissue. Widespread damage to leaves can cause them to drop off. Not so easily observable is a decreasing number of shoots, flowers and fruit. Eventually, if enough of a plant is damaged, it will die.

The most sensitive plants are affected by sulphur dioxide concentrations as low as 0.05 ppm, although experiments with moderately sensitive species indicate the long term threshold level to be 0.15 ppm, with 0.3 ppm for 8 hour exposure (Fig. 3.5). The moderately sensitive species, however, include many common commercial plants – wheat, lucerne, clover, beans, lettuce, fruit trees, etc. – while the hardier species such as certain trees – The Australian She-Oak (Casuarina) for example – are virtually immune. Some species of She-Oak seedlings are quite unaffected by exposures of 6 hours to 3 ppm sulphur dioxide under high sensitivity conditions. The age of the plant at the time of exposure also plays a part. Older plants tend to be more immune to sulphur dioxide than younger ones, but for ozone, chlorine, hydrochloric acid or ammonia, the most sensitive period appears to be just after the plant reaches maturity.

Fig. 3.5 Chronic plant damage as a result of long term exposure to low levels of sulphur dioxide. (From Jacobson, J. S. and Hill, A. C. (eds.) *Recognition of Air Pollution Injury to Vegetation: A Pictorial Atlas,* Air Pollution Control Association, Pittsburgh, Pennsylvania (1970).)

The long term effect of sulphur dioxide on a plant community can be quite significant, although not always as spectacular as the denudation at Queenstown (Tasmania) or Ducktown (Tennessee). A detailed survey of plants starting at a distance of 50 km from the smelter at Wawa, Ontario, showed that 25 species found in virgin, unspoiled areas were present up to within 16 km of the smelter, but from that point on the number of species decreased until none grew closer than 2.6 km, demonstrating that the different species have varying sensitivity. Air pollutants were also affecting new growth – the seedlings – much further away. Some types were found no closer than 25 km, and one type was not found within 48 km. As a result, while the forest itself would grow to within a certain distance of the smelter, its character would be changed over the years.

Ozone

Ozone, one of the major components of photochemical smog, persists over a considerable part of the day. Peak levels in some American cities can exceed the threshold levels (0.1 ppm for 2 hours or 0.03 ppm for 4 hours) for sensitive species. Ozone passes into the plant leaf and attacks the palisade cells (see Fig. 3.4). As a result, the chlorophyll is destroyed, the rate of photosynthesis is lowered, and the respiration rate is affected. The upper cells tend to remain intact and appear healthy, although the inner cells are seriously damaged. The damage finally appears as dark coloured dots or discoloured areas. When held against sunlight the lesions appear red, purple, black or brown, or somewhat lighter in colour. Serious burning of the tips of the needles of the current season is observed in sensitive pines. In geraniums and carnations, ozone prevents (or lowers) the amount of side branching, as well as general growth and flowering. Tobacco plants are an example of a commercial crop which shows early visible signs of ozone damage. This was previously referred to as 'weather fleck' and only in 1959 was it recognized as ozone attack.

Pine forests, and in particular the species of Ponderosa pine in the San Bernardino National Forest in Southern California, are seriously affected by oxidant (principally ozone) pollution. In 1969 it was shown that 1.3 million Ponderosa and Jeffrey pine, which replace the Ponderosa above 2000 m, were affected. Detailed study of some of the Ponderosa pines in this area in subsequent years showed considerable increase in the degree of damage to trees that had, initially, been only slightly affected. Bark beetles were subsequently responsible for the death of the weakened trees. The trees were subjected to comparatively high oxidant concentrations (in excess of 0.1 ppm for about 10 hours daily) during the summer months, with lesser times and lower concentrations at other times.

There is also some evidence that mixtures of ozone and sulphur dioxide can act synergistically – i.e. in concert – and injure sensitive species at concentrations of 0.02 and 0.03 ppm ozone together with 0.02 and 0.10 ppm sulphur dioxide, which are far lower levels than the thresholds for the individual pollutants.

Nitrogen Dioxide
This is similar to sulphur dioxide. It is readily soluble in water, and tends to attack recently matured leaves. Nitrogen dioxide affects plants on dull or cloudy days more than on bright days, which may be because light produces an enzyme reaction in the plant, reducing the nitrite to ammonia, which is a nutrient. Poor light suppresses this reaction. This is also the reason that nitrogen dioxide is less toxic than sulphur dioxide, the threshold for 4 hour exposure being 2.5 ppm, which is not a concentration that would be expected to be found, even in the most highly polluted city air.

PAN
PAN is perhaps the most toxic component in photochemical smog, and

is part of the total oxidants measured. It tends to attack young ornamental plants, vegetables and grasses, rather than shrubs and trees, and acts at much lower concentrations than ozone, 0.01 ppm (for 6 hours) being the threshold concentration. PAN attacks young plants on the underside of the leaf, which tends to have a glazed surface after the tissue below the lower epidermis has been killed. This causes subsequent distortion in the leaf as the upper part continues to grow. Damage to grasses is sometimes seen as dark brown bands.

Fluorides

Fluorides, and in particular hydrogen fluoride, can damage plants in exceptionally low concentrations of the order of 0.001 ppm. The gas is easily soluble and 'acid burns' sensitive tissue, damaging the cells. The leaves then turn brown or light tan and after some days holes develop. Some species of pine and fir are very sensitive and their needles change colour as the fluorides accumulate.

Fluorides also accumulate in grasses and vegetables. These plants, although apparently healthy, can still contain enough fluoride to be toxic to ruminant farm animals, particularly cattle.

Heavy Metals

The heavy metals – lead, cadmium and zinc – all affect plant growth. This is particularly noticeable in the areas surrounding non-ferrous smelters. Zinc and cadmium are more toxic than lead, in that order. The metals are usually in the form of the sulphate ($ZnSO_4$, $CdSO_4$) and are readily absorbed by the plants. Fruit trees are especially sensitive to zinc, and plums are more affected than apples or pears. The type of soil in which the plants grow also has some influence, as some types retain the heavy metal compounds more than others. Some varieties of plant are more resistant than others, and it has been found that the walls of the root cells play an important part in the ability of the plant to absorb the metal compound. Even 4 km from a smelter, the source of the heavy metals, it has been found that plants contained many times the naturally occurring levels, and grasses with these amounts of metal cannot be used as fodder.

The use – or excessive use – of insecticides, weedicides and fungicides, can also cause considerable damage if these spread beyond the place of application. The very fine sprays which are used in the application of these materials result in the formation of aerosols, which can carry far beyond the fields or orchards in which they are being used, and so can affect the ecosystem over a much wider area.

Effects on Animals

The possibility of animals consuming air pollutants deposited on, or

stored by plants, has been mentioned. This indirect effect of a
pollutants has been observed for a considerable time. It tends to occur
near smelters treating non-ferrous ores, and near factories, such as
phosphate fertilizer works, brick kilns and aluminium smelters, where
fluorides are emitted and are concentrated in the grasses in
surrounding fields.

When roasting non-ferrous metals not only are the oxides of these
metals emitted, but also arsenic and cadmium oxides which are common impurities in the ores. As early as 1902 the death of 625 sheep, out
of a total of 3500, from arsenic poisoning was reported in an area
within a distance of 40 km from the Anaconda copper smelter in
Montana. More recently (1955), cattle and horses 5 km from lead and
zinc foundries in Germany were found to be fatally affected. In
Sweden in 1954 a steel works spread molybdenum fumes, which
affected cattle grazing 0.8 km away from the source.

Lead compounds from automobile exhausts are deposited near to
roads, although the concentrations found in vegetation are much
smaller than those near smelters, and have so far been measured as
below the accepted threshold for toxicity to animals.

The signs of heavy metal poisoning are diarrhoea, anaemia and
stiffness. For arsenic poisoning the animals display thirst, vomit and
have garlic breath.

Fluorosis is a more widespread problem, which affects ruminants,
particularly dairy cows. The origin of fluorine compounds, principally
as hydrogen fluoride from fertilizer works, brick kilns and aluminium
smelters, has been pointed out. The fluorine compounds, at sub-toxic
levels for the plants, accumulate therein to about 2 to 3 ppm (based on
the dry mass of the grass). The animals feeding off these grasses first
show mottling and dulling of the teeth, give less milk, and then develop
osteofluoritic bone lesions (spurs) which cripple the animals. For this
reason, chronic fluoride poisoning is called 'staggers'.

Other gaseous air pollutants affect animals in much the same way as
they do human beings, although some animals are less, and others
more, sensitive. The effect of air pollutants on domestic animals
during the 'episodes' in Donora and London were carefully observed.
Cats, to the extent of 12 animals in 165 in Donora, were affected by the
pollution, and of these 3 died. The dogs in Donora were even more
sensitive; of 230 animals 36 were affected and 23 died. Poultry kept in
Donora appeared unaffected, but farmers in the surrounding area
reported 40% fatality among young chicks. In London during the 1952
smog, prize cattle were on show. Of these 5 died immediately, 11 more
had to be slaughtered, and 40 more showed signs of illness which, after
death, could be attributed to emphysema and subsequent heart
failure.

Materials

... have a deleterious effect on materials; stone, ...ed glass, fibre material and others. The soiling effect ... is obvious in industrial cities where buildings of light ... ...ones and bricks soon take on the characteristic black ... ...e erosion of the stonework on buildings of great historic and a... ...ural value is very serious indeed. Some of the great cathedrals in Eu ope, notably that in Cologne which is built of sandstone, are showing signs of rapid deterioration. The Acropolis in Athens has crumbled more in the past 40 years than in the previous 2500. Stained glass windows also suffer badly from air pollution. It is considered probable that if air pollution in the cities containing some of the great Gothic (12th to 16th century) cathedrals is not reduced, these may not last another 200 years, even in those cities which have a relatively clean atmosphere.

Other results of air pollution are the faster deterioration of clothing, curtains and wood, the corrosion of metals and the soiling and subsequent cracking of paintwork. Paint may have been applied to protect woodwork and metal besides having a decorative value.

A fairly spectacular effect of hydrogen sulphide air pollution, which can arise from the biological degradation of stagnant waste waters from sources such as abattoirs, is the attack on white (lead) paints. The white lead oxide in these paints is turned into the black lead sulphate. Such an incident started active work in air pollution control in New Zealand, where white houses near a large abattoir in Auckland turned black in a comparatively short time.

The rapid deterioration of rubber acted on by ozone can be easily demonstrated in the laboratory. In Los Angeles the high ambient oxidant levels are estimated to reduce the life of rubber tyres on automobiles by a significant amount.

Silverware tarnishes rapidly due to hydrogen sulphide and sulphur dioxide in industrial cities. Not only does household and decorative silver require much more frequent cleaning, but silver used in electrical contacts in telephone exchanges has to be protected. This may require air conditioning and special atmospheres so that breakdown and frequent replacing of the contacts is avoided. In Rotorua (New Zealand), where the natural hydrogen sulphide concentration is very high, not only is the telephone exchange building conditioned, with the air being taken in through special absorbing filters, but the racks of switchgear with the selectors are further enclosed and provided with filtered air.

The Cost

Most of the effects of air pollution on plants, animals, property and human health result in a direct cost to individuals and to society. Attempts have been made to assess some of these, although some factors such as discomfort in illness or the loss of works of art, cannot be assessed in simple economic terms. In the USA it is estimated that oxidants alone cause 100 to 125 million dollars damage to vegetation each year. Other estimates assess the total air pollution damage to vegetation at between 200 and 500 million dollars. The US Environmental Protection Agency estimates the loss of vegetation at $325 million and to livestock production at $175 million.

The cost of air pollution to individuals living in a highly polluted community in comparison with an average, less polluted one, is illustrated by a study some 15 years ago in two cities of similar size in the United States; Steubenville, Ohio (a steel manufacturing town), and Uniontown, Pennsylvania, where there was no major large pollution producing industry. In Steubenville, the average spent on external house cleaning, laundry, dry cleaning, hair and facial care was $84 per family more than in Uniontown. However, the higher income groups in Steubenville spent relatively more than the lower income groups in that city. A similar study in Newcastle, Australia, gave analogous results, but it was pointed out that the extra costs in cleaning incurred by living in a more polluted suburb closer to the places of employment were counter-balanced by the higher travelling costs for those who lived in a cleaner suburb further away.

The cost of additional health services, lost working time and other health-connected factors, is almost impossible to estimate but some attempts indicate staggering amounts. Similarly, the costs of repairs to buildings, losses of works of art and related aesthetic losses are not really assessable.

It can, however, be stated that the effective control of air pollution has many benefits. The introduction of 'clean air' zones in English cities, particularly London, has not only resulted in a much cleaner and healthier environment and a reduction in costs of cleaning, but also many more clear sunny days.

Such considerations of cost are conceivable within national boundaries, where the source of pollution is internal and the benefits of clean-up accrue to the local population. The situation is not so simple on the global scale, or over extended areas which incorporate several nations.

Global Changes

Pollutants are ultimately removed from the atmosphere by absorption or adsorption at the surface of the earth or, for particles, by deposition at the earth's surface. Inert gases, for which a destruction mechanism exists in the stratosphere, will diffuse upwards and be converted into products which ultimately will also find a sink at the earth's surface. For most pollutants, to the best of our knowledge, the sinks which exist are adequate to cope with any emissions resulting from man's activities and no long term world-wide changes in the atmosphere occur as a result of emissions. Similarly, it has been assumed that the final recipients of these emissions are able, in the main, to cope with their pollution burden from the air and that no unacceptable accumulations in land or water systems follow, except over relatively limited areas close to the source.

There are three important areas associated with pollutant emissions where it is postulated that these generalizations do not hold, namely: the accumulation of carbon dioxide in the atmosphere, the depletion of ozone in the stratosphere, and the accumulation of sulphates and associated increasing acidity in the biosphere. The latter is, strictly speaking, not global in extent but does occur over extended areas far removed from the original source of pollution, and as a result poses a problem on the international scale.

Carbon Dioxide and the Greenhouse Effect
The most discussed of the long term effects is the changing concentration of carbon dioxide. This has been carefully measured at Mauna Loa in the Hawaian Islands for more than a decade. Seasonal variations have shown up as well as a gradual increase in the average value of some 0.7 ppm each year, to reach a present value of about 320 ppm. The seasonal variation can be accounted for by the greater demand for carbon dioxide by plants during the spring and summer when they are growing rapidly, decreasing the amount in the atmosphere. Records of carbon dioxide concentrations over the last century have shown that there has been a gradual increase from about 290 ppm before 1890 to the present value (Fig. 3.6). The introduction of thermal electricity generation and increases in other combustion processes have led to a rapid rise in the consumption of fossil fuels – essentially stored carbon – and the increasing emission of carbon dioxide, to which the changes are usually ascribed. Comparing estimates of this production with the measured increases would imply that about half the carbon dioxide produced by man is not absorbed by the oceans or stored in other ways.

At first glance, this is difficult to understand as the capacity of the oceans to hold carbon dioxide (CO_2) is 60 times the capacity of the atmosphere, and an equilibrium distribution in the ratio of one to sixty of any excess output would be expected. There are two factors in operation which prevent this partitioning from occurring in the time-scale of interest. Firstly, circulation within the oceans is such that only the upper levels are available to exchange with the atmosphere, and secondly there is a buffer effect which acts to resist any change in

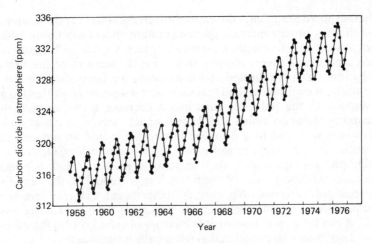

Fig. 3.6 The trend in carbon dioxide concentration at the Mauna Loa Observatory. (Taken from the results of Charles D. Keeling reproduced in *Scientific American*, **238**, 37 (1978).)

concentration within the upper levels. Models of the atmosphere/ ocean exchange have been able to account for the rate of CO_2 increase from fossil fuel burning by taking account of the above mentioned factors. It has since been realized that the forests of the world, which serve as reservoirs of carbon, have been significantly depleted over the last century, thereby releasing carbon dioxide to the atmosphere. The annual emission of CO_2, released as a result of man's forest clearing activities, is estimated to be of the same order of magnitude as the fossil fuel emission. If this is the case the models are no longer adequate and the important question of predicting future CO_2 concentrations remains unresolved. Such information is of considerable importance because of the predicted temperature increase associated with an increase in carbon dioxide concentrations – the so called 'greenhouse effect'.

Carbon dioxide is a strong absorber of light in the infra red region, but is practically transparent to the ultra violet and visible end of the spectrum. This means that the incoming radiation from the sun, predominantly ultra violet and visible, is not obstructed by the carbon dioxide component of the atmosphere while the outgoing infra red radiation from the earth is. The resultant effect is predicted to be a warmer earth atmosphere system accompanied by substantial climatic changes in many parts of the world. The size of the effect depends on the atmospheric model used, and ranges from 1.5°C to almost 6°C for a doubling of the CO_2 concentration. The uncertainty is a function of the present inadequacies of atmospheric models and ignorance about which feedback mechanisms might be significant, for example, whether increasing temperature might increase water vapour content and thereby further increase the temperature effect (water is an infra red absorber), or whether increased water vapour will result in increased cloud cover, thereby increasing the amount of the sun's radiation which is reflected and moderating the temperature effect. Similarly, it might be argued that increased temperature will lead to a reduction in the polar ice caps and a decrease in the amount of incoming radiation which is reflected, or that increased ocean temperatures will lead to a decrease in circulation and an increase in stratification, thereby decreasing the oceans' capacity for CO_2.

Despite our ignorance, the incentive to predict is high because climatic conditions are of such vital importance to the world's agricultural patterns. When it is remembered that the average temperature difference between the present and the last ice age was 4°C, it can be seen that a possible increase of even 1.5°C in the other direction cannot be considered as potentially insignificant.

The trend in temperature which has occurred over the last century cannot be associated with the postulated increase in CO_2 concentration over the same period. Average northern hemisphere temperatures increased by 0.6°C between 1890 and 1940 but since that date have decreased by 0.3°C. This does not rule out the relationship between CO_2 concentrations and temperature but illustrates the uncertainty associated with prediction of climatic change and the magnitude of natural effects of unknown cause.

It has been suggested that particulate concentrations are increasing as a result of man's activities and that this increase would have an effect on the climate. Particles, for instance, increase the albedo (percentage of sunlight reflected) by scattering the incoming radiation, and cooling would be predicted as a consequence. Particles also scatter the outgoing radiation, although not so effectively because the particulate sizes that predominate in the atmosphere are most efficient at scattering light of the incoming wavelengths.

Particles also absorb and re-emit radiation; the wavelengths at which this occurs are a function of the particle composition. In parallel with the greenhouse argument, if this absorption predominated in the infra red region corresponding to the earth's emission, the result would be warming and cooling would ensue on absorption of the sun's emission. Particles, unlike CO_2, have a relatively short lifetime in the atmosphere, of the order of 3 to 5 days. In addition, the man-made contribution to the atmosphere's total particulate burden is relatively small (approximately 10%). It is argued that under these circumstances it is unlikely that there has been any change in particulate concentrations as a result of man's activities, and that there could not be any associated climatic effect. This is not to say that fluctuations in particulate loading as, for example, occur in the stratosphere following volcanic activity, do not have an influence on climate. However, a discussion of natural pollution of this sort lies outside the scope of this text.

Stratosphere Ozone Depletion
The stratospheric ozone layer filters the incoming ultra violet radiation (UV) in the biologically active regions of the spectrum. It is this ozone shadow which made life forms as we know them possible on this earth, and a depletion of the ozone layer can be considered as a threat to these same life forms. The possible outcome of depletion is therefore sufficiently serious to justify intensive study of potential threats, and to counsel caution regarding air emissions which might be contributing to a depletion.

Ozone is formed as a consequence of the interaction of UV radiation from the sun with oxygen to give atomic oxygen (O).

$$O_2 + \text{Sunlight } (hv) \rightarrow 2O \tag{36}$$

This reaction is followed by two others which simply inter-convert the strongly oxidant species and ozone (O_3) without loss.

$$O + O_2 \rightarrow O_3 \tag{37}$$
$$O_3 + hv \rightarrow O_2 + O \tag{38}$$

Ozone is subsequently destroyed by the reaction of atomic oxygen and ozone:

$$O_3 + O \rightarrow 2O_2 \tag{39}$$

The concentration of ozone, in the absence of any other reactions, is a result of an equilibrium between formation and destruction. Equation (39) was found to be insufficient to explain the concentrations of ozone observed and further destruction reactions have had to be postulated. The most important of these is the reaction with the hydroxyl radical OH and the related peroxyhydroxyl species (HO_2), which have already been seen to play such an important role in photochemical smog formation.

$$OH + O_3 \rightarrow HO_2 + O_2 \tag{40}$$
$$HO_2 + O \rightarrow HO + O_2 \tag{41}$$

Overall, this is a loss of one ozone and one atomic oxygen, with no loss of the reactive OH.

The other ozone destruction reaction of importance involves nitrogen oxides, which occur naturally in the stratosphere as a result of chemically inert nitrous oxide (N_2O) released from the soil diffusing into the stratosphere, where it is photochemically converted to the more reactive nitric oxide.

$$N_2O + h\nu \rightarrow NO + N \tag{42}$$

Nitric oxide reacts with O_3 in an analogous way to OH; a loss of two oxidant species without loss of reactant.

$$NO + O_3 \rightarrow NO_2 + O_2 \tag{43}$$

$$NO_2 + O \rightarrow NO + O_2 \tag{44}$$

Equations (36) to (44) are sufficient, in the light of present knowledge, to explain the observed concentration of ozone in the stratosphere. Ozone concentrations are very variable, as much as plus or minus 25% at a particular location on a particular day, leaving an element of uncertainty that all relevant factors have been accounted for. This uncertainty must be reflected in the considerations of how the activities of man can cause depletion of stratospheric ozone. The three important areas where human activity can influence the ozone cycle are: the direct emission of nitrogen oxides by supersonic transport (SST's) flying above the tropopause, additional transport of nitrous oxide as a result of increased use of nitrogenous fertilizers, and the formation of atomic chlorine in the stratosphere from chlorofluoromethanes released in the troposphere.

There is some evidence that increased concentrations of nitric oxide in the stratosphere from the first two sources may not be as serious a threat as first considered. It has been found that the path for stratospheric destruction of nitric oxide is an order of magnitude faster than was once thought.

$$NO + HO_2 \rightarrow OH + NO_2 \tag{45}$$

Increased NO concentrations, lead as a result of this reaction, to a decrease in the peroxyhydroxyl species HO_2 in favour of OH. The HO_2 is a more important ozone destroyer than either OH or NO_2 and it is possible, as a result, that increased NO may actually result in an increase in ozone concentrations. The water vapour from supersonic transports could, in fact, be of more serious concern because it is the source of the OH/OH_2 species in the stratosphere.

Chlorofluoromethanes are used as aerosol propellants and refrigerants because they are inert, and because they are inert they survive in the atmosphere to be transported to the stratosphere, where they are photochemically dissociated to give atomic chlorine (Cl). These chlorine atoms destroy ozone without loss of the reactant species in the same way as OH and NO do.

$$Cl + O_3 \rightarrow ClO + O_2 \tag{46}$$

$$ClO + O \rightarrow Cl + O_2 \tag{47}$$

Predictions of the extent of depletion resultant on continued emissions of chlorofluoromethanes range from 4 to 30%. The uncertainty is a function of assumptions of further usage and ignorance regarding all the relevant reactions and their rates in both the stratosphere and the troposphere. A tropospheric sink, for example, would require a complete reassessment of the predictions, as would the discovery of a significant natural source of chlorinated hydrocarbons sufficiently stable to contribute to stratospheric concentrations of C1.

What would be the effect of a depletion of the order of 15% of the total ozone, which is the World Meterorological Organization's estimate for steady state conditions, based on continued release of chlorofluoromethanes at 1977 levels? The associated increase in UV radiation reaching the surface of the earth could, in some estimates, produce a 30% increase in non-fatal skin cancers although sunlight exposure is strongly related to lifestyle, a hat wearer in a hot climate being less at risk than a sunbather in a temperate one. It is particularly salutary to consider that in some parts of Europe and North America there has been a 200% increase in the UV associated cancers in the past 30 years, presumably associated with lifestyle.

Acid Rain

Increasing acidity in natural waters and soils has become a problem over extended areas of the world, particularly North Eastern America and North Western Europe (Fig. 3.7). This acidity is associated with the transport and subsequent deposition of sulphur dioxide, nitrogen oxides and their acid oxidation products. The concentration levels of the transported gases are not sufficient to damage the environment directly, it is their accumulation which causes the problem. This accumulation was first observed as increasing acidity in Swedish lakes and rivers, an acidity that could not be explained by reference to Sweden's sulphur dioxide emissions alone. It was consequently demonstrated that much of the material responsible had been transported from the highly industrialized areas of the UK and Central Europe, and that this transport had been aided by the tall industrial stacks used in these areas to avoid a local problem. Similar conclusions were drawn from Finland, Norway, Austria and Switzerland, all of which were found to be net importers of sulphur compounds, in the course of an OECD Study during 1973, 1974 and 1975.

Perhaps the most worrying aspect of this sulphur problem is that the area affected by acid precipitation is increasing year by year, and its association with our energy consuming world leaves little prospect of alleviation, even with stricter emission controls.

The repercussions on the environment are difficult to quantify because they are a function of soil type and bedrock. It does seem, however, that the silaceous bedrocks, thin soils and soft waters of the areas at present receiving the deposition are particularly vulnerable. This is related to the relative insolubility of the rock and the con-

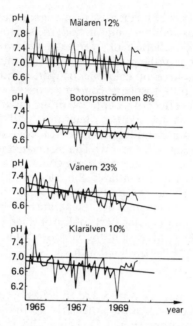

Fig. 3.7 Changing acidity in Swedish rivers and lake outlets. (From Sweden's case study for the United Nations Conference on the Human Environment, Stockholm (1971).)

sequent lack of dissolved species in the water which could counter (buffer) the acidity change. Significant reductions in fish populations have been observed, accompanied by an associated decrease in the variety of species at all levels in the food chain. There are reported at present to be some 15 000 fishless lakes in Sweden as a result of increased acidity, and some 100 such lakes in the Adirondak region of the USA. Soils in these regions similarly have little buffering capacity and are subject to the leaching of minerals essential to plant growth. There are soils which are low in sulphur or high in carbonate, where an increase in acid sulphate could well be benign or even beneficial, but these are not predominant in North Western Europe or North Eastern USA.

4 Measurement of Pollutants

Improving or even preserving air quality requires measurement of pollutant concentrations in the atmosphere. Such measurement serves several interrelated but definable functions:

1 Measurement provides the necessary data on which the relationship of effects to pollutant levels can be determined.
2 Measurement provides a measure of whether air quality standards are being met or exceeded and to what extent.
3 Measurement is necessary to determine whether any deleterious changes in the global background levels of pollutants are occurring as a result of man's activities.
4 Measurement provides the data for determining the fate of pollutants in the atmosphere and, coupled with meteorological information, is an aid to modelling and therefore predicting the relationship between concentrations, emissions and weather conditions.
5 Source measurements provide information about emissions and, where applicable, whether the emitter is meeting the regulatory standard.

In each of these circumstances, the pollutants to be measured, the duration of the measuring programme, the sensitivity of the method required, and the period over which the concentration is averaged, will be different. For example, in choosing which pollutant is to be measured, the group of pollutants that are suspected of having adverse effects on human, animal and plant health and on art and property, are not the group of pollutants that are accumulating in the atmosphere. In the first case, we are interested in the reactive components of emissions whose lifetime in the atmosphere is relatively short, whereas in the second it is the unreactive components that are of interest. Similarly, in a study of dispersion for modelling purposes an unreactive representative pollutant would be chosen for a tracer, whereas the chemistry of photochemical smog would require all the possible reactive species to be measured.

There is an interesting problem associated with choosing the pollutants for measurement of effects, namely that we require a prior knowledge of which pollutants may cause an effect in order to demonstrate which pollutants do cause an effect. This involves an element of trial and error based on experience of past air pollution episodes, laboratory experiments and a knowledge of the major anthropogenic components of urban airsheds. Nevertheless, many problems, including synergistic enhancement of effects and the outcome of long term exposure to elevated but sub-acute levels on human communities, remain unsolved (Chapter 3).

To take as another example the period over which the pollutant concentration is averaged, this need be no shorter than is consistent with the purpose of the study. The closer one approaches real time measurement, the more expensive the equipment involved becomes because of the greater sensitivity required and the amount of data which has to be stored and processed. Averaging time corresponds to the sampling time used, for a collected sample, and to the response time of the instrument for a field instrument with direct response to changes in concentration. The required response time for deter-mination of human health effects is a function of the body's response time to changes in pollutant concentrations in the atmosphere. For a pollutant like sulphur dioxide, this response is fast and short term atmospheric peaks of this species may be the critical factor in determining health effects, whereas for a cumulative pollutant like lead the response is slow and the long term average is all that is necessary. These differences are reflected in air quality standards, where these are operative, as a 30 minute average for the SO_2 standard and a 30 day average for Pb.

Ambient background concentrations of pollutants are often in the parts per billion (ppb) range or less, as we have seen in Chapter 1. These concentrations are a function of a natural cycle of emission and removal that overshadows the anthropogenic one. The changes in back-ground concentrations which are a result of pollutant emissions are, therefore, only a small fraction of the background concentration itself. This requires very sophisticated analytical techniques, free from interferences by any of the many other atmospheric constituents, and sufficiently sensitive to discriminate differences in concentration values at the level required. The requirements for detection limits in urban and industrial situations, where the anthropogenic contribution may predominate, is not as great. The number of possible interferents, on the other hand, is greater and the concentrations of most species of interest are extremely variable. Concentrations at the source are greater still and other problems complicate the measurement, the most frequent being the hot, wet, dirty character of most emissions at their source.

Measurement of Source Concentrations

In the case of some industrial air pollution sources, the monitoring of the waste gases being emitted is an important part of process control, and has been standard practice in large processing plants for many years because lack of control can result in loss of valuable product. For example, high concentrations of carbon monoxide and unburnt carbon particles in the flue gases from boilers represent the loss of valuable heat, while high sulphur dioxide concentrations in sulphuric acid plant waste gases represent a loss of product acid.

The ideal methods of measurement are those which give almost instantaneous results, which can then be used for process control, as well as determining pollutant emissions. Invariably they are physical, as those employing chemical methods even when fully automated take several minutes. Of course, complex controls are not common on small boilers. Adjusting these for optimum combustion requires the use of portable measuring devices. Usually these devices have chemical absorption systems and are low in cost. However, they require some knowledge and considerable skill in operation, which simple physical systems ('black boxes') do not.

In the simple chemical method for analysis of combustion gases (represented by the Orsat method), a sample of flue gas is withdrawn from the flue or chimney, passed through a filter to remove particles, and then passed in turn through sodium hydroxide solution (to absorb carbon dioxide), alkaline pyrogallol solution* (to absorb oxygen), and finally acid cuprous chloride solution* (to absorb carbon monoxide). Sulphur dioxide is absorbed together with carbon dioxide. To determine sulphur dioxide in flue gases, a separate sample has to be taken and a method of analysis independent of carbon dioxide, such as iodimetric titration, used.

Physical methods used for these gases are infra red spectroscopy for carbon dioxide, carbon monoxide and sulphur dioxide, and ultra violet spectroscopy for sulphur dioxide, nitrogen dioxide, and others. A simple method for carbon dioxide uses differential density (relative molecular mass of $CO_2 = 44$ compared to $O_2 = 32$ and $N_2 = 28$), or thermal conductivity.

Particulate matter coming from boiler and furnace chimneys can be assessed with the classical Ringelman chart. The smoke density is compared with the fractional darkening (from 20 to 80%) of the paper, representing the 'Ringelman Numbers' from 1 to 4 (Fig. 4.1). This can be automated by measuring the obscuration of a beam of light shining across the chimney onto a photoelectric cell. It is also possible to sample the particulate matter passing up a chimney, and then chemically determine its composition as well as measuring particle concentration, number and size by microscopic or other means

*Many alternative reagents are mentioned in the literature which have great merit, particularly as an absorbent for carbon monoxide. However, these are the most common and easiest to prepare.

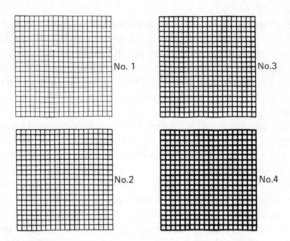

Fig. 4.1 Standard Ringleman chart. (From Parker, A. (ed.) *Industrial Air Pollution*, McGraw-Hill (UK) (1978).)

A very important aspect of the sampling techniques is the question of how representative the sample taken is of the source. The final result of any analysis will only correspond to the true value if the sample taken was a representative one. In stacks (flues and chimneys), emissions can vary in time as changes occur in the process in the plant. They can also vary in space within the stack as a result of wall interactions and the effects on flow of obstructions and bends within the stack. These factors must be taken into account when samples are being taken, and generally involve sampling at several points across two diameters of a stack at right angles to each other. When particulates are being sampled, the rate of sampling must be the same as the rate of flow within the stack or some particle sizes will be sampled preferentially to others.

Automobile emissions are usually taken from the exhaust of a vehicle when it is operating in a stationary setting on a brake dynamometer. The dynamometer simulates a predetermined driving cycle for the vehicle, representative of the average cycle for a car on the road. A constant proportion of the exhaust gases are then collected in a plastic bag, together with dry clean air or nitrogen, for final analysis in the laboratory. The make up air or nitrogen is necessary to prevent the condensation of water in the cooling exhaust gases, which could serve as centres for reaction, absorb gases from the exhaust, or even condense on particulates in the exhaust. Alternatively, gases can be analysed continuously and the results obtained for emissions under different driving conditions.

Testing of automobile emissions is increasingly becoming part of the regulatory requirement for road worthiness in countries like the USA where the vehicle population is high and the conditions for photo-chemical smog formation favourable.

Measurement of Ambient Concentrations

Until recently, it was only possible to measure the average amounts of pollutants in the air over a period of time, and from these measurements establish long term trends in pollution levels. It is now practice in some industrial areas, notably The Netherlands and Japan, to use virtually instantaneous measurements drawn continuously from a network of measuring stations ('real time monitoring') to modify industrial activity and reduce high pollution levels under adverse weather conditions. Nonetheless, the classical methods of taking samples over a period of time to assess air pollution continue to be used because, as has already been discussed, real time measurements are more expensive and not always necessary, and thus classical methods remain the base method against which newer ones are evaluated. Sampling followed by analysis is, in addition, the only satisfactory method of determining particulate composition and of analysing atmospheric samples for individual hydrocarbons.

Particulate Matter

The largest of the airborne particles can be collected as they fall out of the atmosphere, usually in a cylindrical vessel of standard dimensions with a funnel on top and containing added water to prevent re-entrainment of the particulate. This method does not give results which are representative of particle concentrations in the atmosphere because only the largest particles ($> 20 \mu$m) are heavy enough to be deposited in this way without being affected by wind movements. The total mass of particles collected is a function of collector design, and comparisons are only useful between collectors of identical design which have been used in an identical way (i.e., located at the same height above a similar surface). In its favour, the method is cheap and requires little attention, samples only being collected and weighed every 30 days. The method's main use lies in studying trends over a long period of time and in studying the distributions in emissions from a single source under varying conditions. Because dustfall was the earliest of the particle collecting techniques to be standardized, there is data available over a longer period of time, giving a measure of trends in air quality at the sites where the collectors were located. Results are generally reported as mass per unit area per month. Extrapolation of these results to give total fallout should be avoided because of the difficulties in interpreting the results in absolute terms.

Another frequently used method of determining particulate levels is to draw the contaminated air through a filter and measure the degree of soiling, either by measuring the decrease in transmittance of light of the filter (Coefficient of Haze, COH) or the decrease in reflectance (Reflectance Units of Dirt Shade, RUDS). Obviously, these are not absolute measures of the level of contamination of the air as both COH and RUDS will depend on the properties of the dust collected, different sources giving dusts of different reflection and absorption characteristics. These measures have been found to correlate reasonably well with minor childhood respiratory complaints in industrial European cities. They have the further advantage of not requiring a large sample and can, therefore, be conveniently used to sample sequentially with time using a moving filter tape. In areas where the pollution sources are invarient with season and time of day, it has been possible to convert COH and RUDS values to concentration of particulate in units of mass per unit volume.

Atmospheric mass per unit volume concentrations are generally obtained using a 'high volume' sampler which pulls a known volume of air through a filter (Fig. 4.2). The high efficiency filters used collect almost 100% of the dust above 0.1 or 0.2 μm in diameter. High volume samplers with high air flows, 1-2 m^3 per minute, and large air inlets, 150 by 200 mm, collect a sample of atmospheric particulates whose representativeness is largely unaffected by meteorological conditions.

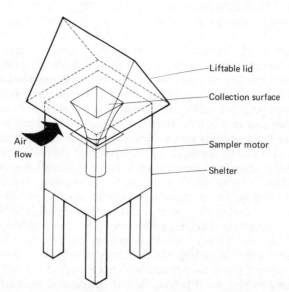

Fig. 4.2   High volume sampler in a standard shelter. The shelter serves to reduce the collection of particles above 50 μm.

Concentration of particulates is obtained by weighing the filter at constant humidity, both before and after collection, and dividing the change in mass by the total air flow.

Samplers are also available which separate particles into different size fractions by making use of the aerodynamic properties of particles of different sizes. This size differentiation parallels what happens in the human lung (Chapter 3) and is important in determining what percentage of particles in the air could potentially be deposited deep within the lung, where they or their chemical components could cause damage or be absorbed into the blood stream.

A variety of analytical techniques are available for determining the concentration of components of interest in collected particles. Metal ions, such as lead, cadmium, mercury and zinc are most commonly determined by atomic absorption after acid digestion to extract the metals of interest. Any technique capable of measuring metal ions in the required concentration range could be used provided the possible interferences present have been accounted for.

Organic components of aerosols are determined by first obtaining the sample in solution using solvent extraction techniques which do not decompose the compounds of interest, followed by standard analytical techniques for trace concentrations, generally a chromatographic separation of some sort – thin layer chromatography, liquid chromatography, or gas chromatography. Detection of the separated compounds can be by ultra violet spectroscopy, which is suitable for most organics, or fluorescent spectroscopy, which is more sensitive but only useful for compounds that fluoresce in ultra violet light. This last group includes the potentially carcinogenic polyaromatic hydrocarbons discussed earlier. Complete identification requires the use of mass spectroscopy but this is not always necessary when pure standards are available for comparison with unknowns in the chromatographic separation step.

Gaseous Pollutants
Gaseous pollutants can be quantified in the atmosphere in three main ways:

1   A sample can be collected in an evacuate vessel or a plastic bag and transported back to the laboratory for analysis (grab sampling). This method gives the concentration of pollutant in the atmosphere at the instant the sample was taken, provided the sample was a representative one. The time delay between taking the grab sample and making the actual measurement can result in losses of sample by preferential absorption on the surface of the container, or by preferential diffusion out of the container in the case of plastic bags. These problems can be reduced by preconditioning the containers with

the atmosphere being sampled, using plastics which are known to be impermeable to the gas of interest and by carrying out the analysis as soon as possible after sampling.

2   Samples can be collected and concentrated at the same time by preferential absorption onto solid adsorbants, absorption into solution, freezing out components at selected temperatures, or some combination of all these three. Samples are then returned to the laboratory for analysis. Problems here can occur as a result of concentrating reactive pollutants, increasing their chance of reaction before analysis. For gas phase organic species, which are not reactive in the dark, both cryogenic methods and the use of solid adsorbants have proved to be very successful. The organics can then be analysed by the same chromatographic methods discussed earlier for particle phase organic species.

The earliest methods involved the collection of gaseous pollutants by absorption into solution. This is generally done by drawing the air at a known rate, first through a filter to remove particulates, and then through the absorption solution held in a gas collection bottle designed to ensure that the gas bubbling through it presents the maximum surface area to the solution. In many cases, the solution chosen has an irreversible reaction with the gas pollutant, thereby fixing it in solution. For sulphur dioxide, nitrogen oxides and oxidants, these methods remain the standard against which new methods are compared.

All collection with concentration methods require finite sampling times whose length depends on the concentration of the species in the atmosphere. They therefore provide an average concentration over the length of time of the sampling. Again, as with stack sampling, care should be taken to ensure the sample taken is representative of the bulk from which it came.

3   Finally, concentrations can be determined continuously using a physical property of the gas being measured, or a physical property of the product of a specific controlled reaction of the gas being measured.

The application of these general approaches to specific cases will now be considered.

Chemisorption
Sulphur dioxide has been measured for over 40 years in England by the 'lead peroxide candle' method. This is basically a chemisorption (absorption with chemical reaction) technique, involving the reaction of lead peroxide with sulphur dioxide:

$$SO_2 + PbO_2 \rightarrow PbSO_4$$

The lead peroxide is coated as a paste on a ceramic support of candle shape, which must be protected from the weather. The total amount of sulphate formed over the time of exposure (usually 30 days) provides a measure of the integrated sulphur dioxide concentration over that period of time. The method is cheap and requires no special skills. However, the collection efficiency is a function of humidity, temperature and wind speed and therefore only satisfactory for order of magnitude determinations.

Another adaptation of the chemisorption technique is the use of chemically impregnated crystals, which undergo a colour change as a result of reaction with the compound being measured. Tubes containing specific reactants for a wide range of pollutants are available commercially. The tubes are opened at the test site and a known volume of air is drawn with a hand pump through the absorbant. The distance along the tube that the characteristic colour change occurs provides a measure of the concentration. Such methods are most useful in industrial plants where concentrations are higher than is normal in ambient air.

Wet Chemical Methods
The most commonly used non-instrumental methods for the reactive gases are those involving collection in solution with reaction, referred to as wet chemical methods or manual chemical methods.

For sulphur dioxide, the commonly used wet methods are the West Gaeke and the hydrogen peroxide methods. In one version of the West Gaeke, which is a modification of the original procedure suggested by West and Gaeke in 1956, the air sample is collected in potassium tetrachloromercurate solution. After a predetermined sampling time, the solution is analysed by adding pararosaniline and formaldehyde to form the intensely coloured pararosaniline methyl sulphonic acid, which can then be measured spectrophotometrically. Hydrogen peroxide collection leads to oxidation of the SO_2 to sulphate (SO_4^{2-}), which is subsequently precipitated as barium sulphate in the laboratory. Barium sulphate remains in suspension and can be quantified using light attenuation in a spectrophotometer.

The wet methods generally used for nitrogen oxides are modifications of the Griess-Saltzmann method. This involves collection in an absorbing solution containing sulphanilic acid and a naphthylamine, which gives a pink azo dye complex with hydrolysed nitrogen dioxide, NO_2. NO_2 can then be determined spectrophotometrically as the azo dye. Nitric oxide, NO, can be determined by passing a parallel stream of air through an oxidation tube containing supported chromic oxide to convert the NO to NO_2, prior to collection and conversion to the azo dye. The nitric oxide concentration can then be obtained by difference. Modified Griess-Saltzmann methods have not been established as reliable, in particular the stoichiometry is not what is theoretically expected, and there are problems in some modifications with collection efficiencies. Results from such measurements must, therefore, be treated with some caution.

Total oxidants are determined manually using buffered potassium iodide as the collection medium. Air, after filtering, is passed through neutral potassium iodide solution (KI) in a phosphate buffer, where the following reduction reaction takes place:

$$O_3 + 2H^+ + 2I^- \rightarrow I_2 + H_2O + O_2$$
$$I_2 + I^- \rightarrow I_3^-$$

The tri-iodide ion (I_3^-) is then measured colorimetrically. The method responds to all oxidants although not at the same rate or to the same extent. For example, the response to ozone and peracids is fast but to other peroxides slow, and the response to NO_2 is only 10% of the response to ozone. The value obtained is therefore a complex composite of all the oxidants present.

Monitoring Methods

The earliest automatic monitoring instruments were adopted from the wet chemical methods previously described. A sample was collected over a preset length of time, a measurement was made, and the reagents automatically renewed for the taking of another sample. One of the main problems with wet methods for automatic operation is the inherent problem associated with storage of large volumes of solution, which may be unstable or evaporate and thereby change in concentration and, of course, have to be frequently renewed. Averaging time for each species is dictated by the sampling time required to give a significant instrumental response, and for the reactive pollutants this time proved to be too long for many applications.

The greatest success to date with continuous monitoring methods has been with instrumentation which measures some physical property of the gas of interest, or a physical property of a product of a gas phase reaction of that gas. Such instruments are characteristically specific as well as having added advantage over the wet methods for sensitivity, reliability and low manpower requirements. Some of the most common of these instruments in present use are outlined in the following sections.

Carbon Monoxide

Carbon monoxide is a strong absorber in the infra red (IR) portion of the spectrum and this property is made use of in the non-dispersive IR instrument (Fig. 4.3). This instrument consists of an infra red heat source, a sample cell through which the sample is pumped, and a reference cell containing a non-absorbing gas. The detector is a two

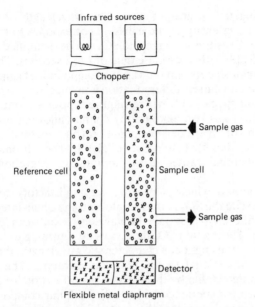

Fig. 4.3 Non-dispersive infra red gas analyser for carbon dioxide analysis. (From Butler, J. D. *Air Pollution Chemistry*, Academic Press (1979).)

chamber vessel with the chambers separated by a pressure sensitive diaphragm and containing equal concentrations of carbon monoxide. The emitted radiation passing through the sample cell will be partly absorbed by the CO in the sample, leading to a lower absorption on the sample side of the detector than on the reference side. The reference side of the detector will be warmer and, therefore, exert a greater pressure on the diaphragm. A chopper is introduced because a pulse in the diaphragm can be more accurately converted into an electrical signal than an absolute deflection.

All infra red absorbing species in the air sample will interfere to some extent, particularly CO_2, hydrocarbons and water. At normal atmospheric concentrations, CO_2 and hydrocarbons are not a problem and can be adequately accounted for in the reference cell gas. Water causes more problems, being both a strong IR absorber and very variable. One solution is to pre-treat the incoming air to a constant water content, or alternatively to filter out all those carbon monoxide absorption bands that are subject to interference by water.

Sulphur Dioxide

Sulphur compounds, when introduced into a hydrogen rich flame, are reduced to diatomic sulphur, S_2, which is excited in the hot flame. The excited diatomic sulphur emits light at a specific wavelength (375 nm) when returning to the ground state. This wavelength is specific to sulphur and the intensity of the emission corresponds to the sulphur concentration. Sensitivity is high and SO_2 can be quantified down to a minimum of 5 ppb with a response time of 2 to 3 seconds. The method does not distinguish between the various compounds of sulphur in the air. In most circumstances differentiation is not necessary because SO_2 is the dominant species but, where required, prior separation can be carried out on a gas chromatography column allowing for separate determination of hydrogen sulphide (H_2S), methyl mercaptan (CH_3SH), and dimethyl sulphide ($(CH_3)_2S$). The chromatography requires sequential rather than continuous operation of the instrument.

Another method which is proving very satisfactory for sulphur dioxide is to excite the SO_2 in the sample air using a high intensity ultra violet source filtered to pass only the wavelengths between 230 and 190 nm (Fig. 4.4). The excited SO_2 fluoresces on returning to the ground state, that is, it emits light at a characteristic wavelength, the intensity of which is proportional to the SO_2 concentration. The minimum concentration detectable by this method is of the order of 2 ppb and no interference is experienced from other sulphur compounds. The main problem is to ensure that the excited SO_2 is not quenched before emission, and this is done by keeping the pathlength of the emitted light short. Self-quenching by absorption by another SO_2 of the emitted light puts an upper limit on the concentration range of 200 ppb.

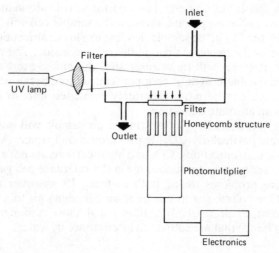

Fig. 4.4 Fluorescence sulphur dioxide analyser. (From Strauss, W. (ed.) *Air Pollution Control – Part III*, Wiley Interscience (1979).)

Nitrogen Oxides

Nitric oxide reacts with ozone (O_3) to give an excited nitrogen dioxide which emits light on returning to the ground state. This phenomena is called chemiluminescence because the emission is induced by a chemical reaction. Like the previous light emission methods, it is specific to NO and very sensitive, down to 1 ppb (Fig. 4.5). The ozone is produced in a spark generator and in this instrument the problem of quenching is handled by maintaining the reaction vessel at a reduced pressure. A filter ensures that only light in the wavelengths of interest is measured. Total nitrogen oxides (NO_x) can be obtained by conversion of NO_2 to NO in a hot (650°C) stainless steel tube.

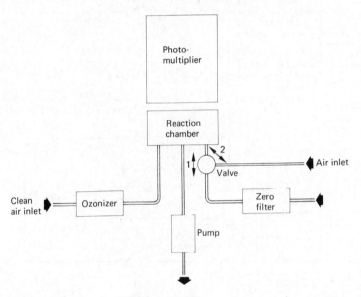

Fig. 4.5   Chemiluminescence nitric oxide analyser. Nitrogen dioxide is measured by subtracting the nitric oxide value from the total nitrogen oxides figure obtained by a preliminary conversion of nitrogen dioxide to nitric oxide. (From Strauss, W. (ed.) *Air Pollution Control – Part III*, Wiley Interscience (1979).)

Ozone

Ozone is also monitored by a chemiluminescence method, in this case involving the reaction between ozone and ethylene which results in an emission at 435 nm. The lower detection limit for ozone is 1 ppb, well below the background concentrations normally found of 10 to 70 ppb. Nitrogen oxides do not interfere.

Hydrocarbons

Hydrocarbons are monitored using a flame ionization detector (FID).

This consists of an air/hydrogen flame across which an electric potential is passed. When hydrocarbons are pyrolysed in the flame, its conductivity greatly increases and a current passes which is proportional to the number of carbon atoms present. Commercial instruments generally include a means of separately measuring methane, carbon monoxide and total hydrocarbons; the concentration of non-methane hydrocarbons which is generally of greatest interest can then be found by difference. The three way determination requires sequential operation (Fig. 4.6). The air sample is introduced directly

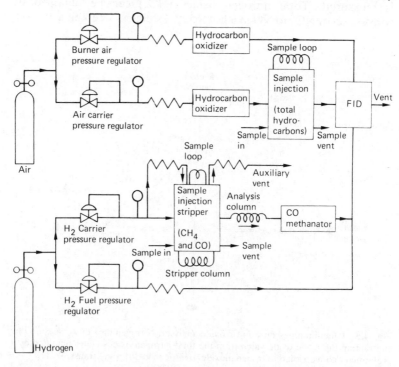

Fig. 4.6 Three component system for methane, non-methane hydrocarbons and carbon monoxide. Beckman model 6800 taken from Perry, R. and Young R. (eds.) *Handbook of Air Pollution Analysis*, Chapman and Hall (1977).

into the flame to give the total hydrocarbon value. A second sample is passed through a gas chromatography column for a time sufficient to allow only methane and carbon monoxide to pass. Methane and CO are then separated from each other by a further column and passed through a catalytic methanator to convert the CO to methane before being passed into the FID. Another approach to obtaining non-methane hydrocarbons is by controlling the flame to burn all hydrocarbons except methane.

Calibration

All analytical methods for determining trace concentration must be calibrated against standards. Such standards need to be exact, reproducible and of known concentration, as they form the basis for relating instrument response to pollutant level. The standard should be in the same matrix (generally air) and within the same concentration range as expected in the field. In addition, concentrations should be known independently of any chemical measurement. This can be done by either diluting pure gases to a known extent in air or nitrogen, or by making some physical measurement on the calibration standard. Because of the low concentrations required, the diluent air or nitrogen must be particularly pure. Standard gas mixtures are also necessary for determining the collection efficiency of wet collection methods, for quantifying losses in grab sampling techniques, and to provide test mixtures for laboratory effects studies.

Monitoring Networks

Real time monitoring devices as discussed in the previous section have led to the development of monitoring networks in urban areas with pollution problems. The prime function of these networks is to act as an early warning system of conditions that might lead to unacceptable levels of pollutants. They consist of a number of monitoring stations, each containing a range of pollution monitoring and meteorological instrumentation and all connected to a central computer where the data is processed and interpreted. Such systems are in operation in several cities in Japan, Holland and Germany.

Networks provide information which can be used to detect immediately whether an accidental or illegal emission has occurred in any part of the monitored area, enabling fast corrective steps to be taken. In some cities in Japan, emission data from major industrial plants is also fed into the central computer, enabling the compliance of individual factories to regulatory requirements and directives to be monitored. The most important reason for monitoring, however, is to provide the data for predicting the onset of unacceptable concentration levels so that appropriate preventative measures can be taken. Predictions are based on past experience of the association between concentrations and meteorological conditions, together with a knowledge of the emission pattern for the region. In the ideal case, this past experience would be formulated in a model, validated for the airshed, which can then be used with confidence to make reliable predictions. As yet, such models as exist, although useful, are not entirely reliable particularly for a complex reactive system like photochemical smog; nor are they universal in the sense that a model developed for one airshed can be transferred to another.

The type of action which can be taken when unacceptable levels are predicted depends on the pollutant and its source. For example, high levels of photochemical smog precursors on days of high sunlight intensity in Los Angeles lead to appeals to individuals not to use their automobiles. In the Rijmond area of Holland, where industry is a predominant supplier of photochemical smog precursors, similar conditions lead to a request to industry to cut down emissions by 20%. Rijmond also has an alert for high sulphur dioxide levels (greater than 0.12 ppm). At these levels, industry is requested to change to low sulphur fuels. Such high levels occur under stagnant conditions and SO_2 build-up is taken as an indicator that such stagnation has occurred. An example of a monitoring network, that of the Rijmond area, is shown in Fig. 4.7. A total of 31 stations cover an area of about 80 km² (representing one sampling site in each 2.6 km²). The density of sites required is a function of the complexity of the terrain, the population affected, and the range and level of emissions. Each measuring site transmits the reading on the instrument every 64 seconds by a frequency modulated signal to the central station, where a computer integrates the measurements for each site over each hour. The measurements are then coupled with meteorological data (wind speed, direction, temperature, etc.) for a period of several hours and trends in pollutant concentrations are predicted. These trends indicate whether any action will be necessary.

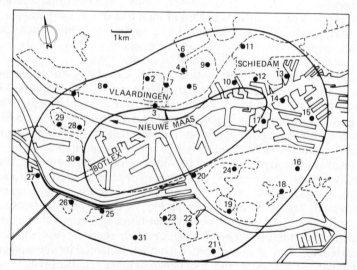

Fig. 4.7 Monitoring sites in the Rijmond area of the Netherlands. The inner circle encompasses the highest industrial density. (From Strauss, W. (ed.) *Air Pollution Control – Part III*, Wiley Interscience (1979).)

5 Air Pollution Control

Air pollutants, even at their source, are usually present in quite low concentrations in large volumes of an inert carrier gas. After dispersion in the atmosphere they are further diluted, and so it is essential that they are controlled before emission at, or as close as possible to, the source. This means treatment at the plant rather than at some central depot, as can occur with liquid waste. The exception is when there are a number of small similar sources grouped together, such as wood working or metal grinding machinery. It also means that mobile pollution sources, such as automobiles, trucks or buses have to carry their pollution control systems with them. Vast numbers of these are therefore required in the case of automobiles, and they have to be particularly reliable in operation as they can only be checked infrequently.

One method of controlling air pollution which should always be considered is changing the method of plant operation, or changing the raw materials used in the process, to remove or reduce pollution. For example, using gas as a combustion fuel instead of coal or distillate, or changing the combustion process to produce less CO or NO_x. A variation on the change of combustion fuel approach is to 'purify' the fuel, to remove its pollutant producing constituent before it goes into the industrial boiler. The desulphurization and de-ashing of coals is an important example of this approach, particularly as coals are likely to be increasingly used as a source of energy in the future as supplies of gas and liquid petroleum products decline.

If the control measures economically available are insufficient to reduce atmospheric concentrations to 'safe' levels, then it may be necessary to stop production altogether. In certain circumstances, the added burden of industry in urban areas causes concentrations of pollutants to exceed acceptable limits; in these circumstances re-location may be the answer. Thus, electricity, which was at one time generated mainly in cities, is now more often made at power stations located close to the fuel supply. Similarly, chemical plants and heavy industry are grouped in industrial areas away from concentrations of population to minimize the nuisance they cause. One problem that arises out of zoning procedures of this sort is that prosperous industry attracts people, and it becomes difficult to prevent the encroachment of residential and commercial areas on what was intended to be a 'green belt' or 'sanitary buffer zone' between residence and industry. Thus, air pollution control is not only a matter of the technical control of the processes producing the air pollutants, but it involves a complex pattern of economic factors, urban planning, and legislative controls.

The sources of air pollution produce gaseous mixtures, fine particles, or both. The classical form of air pollution is smoke – fine particles of carbonaceous material from incomplete combustion – and ash, often from poor combustion and overloaded equipment. This can be partially, and often adequately, controlled by better design of the combustion equipment, careful adjustment of burners, and avoidance of overload conditions caused, for example, by sudden changes in demand*. Other possibilities are changing of fuel or type of combustion system.

In the United Kingdom, smokeless 'clean air' zones have been proclaimed in many cities, notably London. Now, only special smokeless fuels can be burned in open fireplaces, otherwise smokeless appliances (or smoke consuming appliances) have to be used. This is in addition to very strict control of industrial plant, combustion units, furnaces and boilers. As a result, London has become a much cleaner city, and it is claimed that the days with sunshine and clear skies have markedly increased in the past 20 years (Fig. 5.1).

Smoke from domestic heating and cooking remains a serious problem in the less developed regions of the world, where much of the population has neither the technical sophistication nor the economic

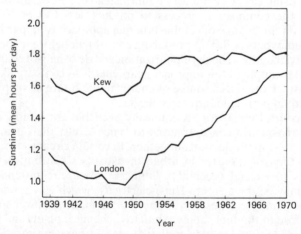

Fig. 5.1 Winter sunshine trends in London and at Kew. (From Holdgate, M. W. *A Perspective of Environmental Pollution*, Cambridge University Press (1979).)

*In Zürich, Switzerland, thousands of oil fired household boilers have been adjusted or modified by the municipal authorities in their campaign to reduce pollution.

resources for smokeless fuels or smoke consuming appliances. However, in the developed countries our major air pollution problems are associated not with particles but with gaseous pollutants for which we do not always have a control technology, able to operate on the scale required within acceptable economic limits.

Removal of pollutants from an emission represents two quite separate problems, depending on whether the pollutant is gaseous or particulate. Whereas particles can be acted on by a whole range of physical forces including gravitational and electrostatic, gases are homogeneous with the carrier gas and can only be removed by diffusion to controlled surfaces where they are preferentially adsorbed or absorbed or, in some cases, chemically altered to a less noxious product.

Control of Gaseous Pollutants

For gaseous pollutants to be controlled they must be removed from their gaseous environment to either a liquid or solid surface, where they are preferentially retained, or where they react to form a non-polluted species or a species that is more readily removed than the original contaminant. The processes used are variations of the methods used for collecting gases with concentration, i.e., absorption into a liquid (scrubbing) and adsorption onto a solid surface either with or without reaction.

Gas Absorption in Liquids

Gas absorption in a liquid, which occurs in the scrubbing process, is a standard chemical engineering unit operation, technically developed and relatively well understood. When dealing with comparatively high concentrations of a contaminating gas (of the order of 1% or higher), it is frequent practice to use a 'counter current' flow system in a unit such as a packed absorption tower, shown in Fig. 5.2. This has the advantage that the lowest concentration of the pollutant in the gas is in contact with the 'weakest' liquid, which is the absorbent liquid in which there is the least concentration of the contaminating gas (or perhaps pure absorbent liquid if it is not recirculated in a closed system). The most concentrated liquid leaving the absorption column contacts the highest concentration of contaminant. The used absorbent liquid can then be run to waste or treated so that it can be recycled. In many cases, the 'contaminant', when removed ('stripped') from the absorbent liquid, can be used as the basic material for further processing. Thus, a common operation in petroleum refineries is the absorption of hydrogen sulphide in an alkaline solution, its subsequent stripping using steam, and then converting into sulphur. This sulphur is the raw material in fertilizer (superphosphate) production.

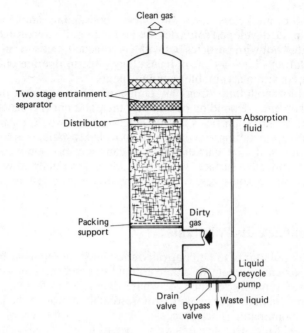

Fig. 5.2 Diagram of a packed absorption tower for gas cleaning. (From Strauss, W. *Industrial Gas Cleaning*, Pergamon Press (1966).)

Air pollutants which are present in very low concentrations are often scrubbed in a 'co-current' system, with the gas following in the same direction as the scrubbing liquid.

The size of the scrubber and its relative effectiveness is a function of a number of parameters: the surface area of the absorbing liquid, the temperature, the time available, and the 'driving force' which takes the pollutant gas molecules to the surface of the liquid and affects their absorption. This depends on the chemical nature and the interaction of the pollutant gas and the liquid. If the gas is easily absorbed, for example ammonia in water, this has a large driving force and the system required can be relatively smaller than in the case of a difficult absorption, such as sulphur dioxide in water. The normal choice for SO_2 scrubbing would not be water but an alkaline solution, such as ammonia, in which SO_2 is very soluble. The driving force is also a function of the concentration of the pollutant gas in the gas stream and at the liquid surface, slowing down as the liquid approaches saturation with the gas.

The surface area through which a gas is absorbed is a function of droplet size of liquid (in a spray type scrubber) or size and type of packing (in a packed tower), as well as the amount of liquid per unit volume of gas used and the physical dimensions of the system. The calculation of the size of an absorption tower to achieve a given degree of gas purification is a standard chemical engineering calculation. It should, however, be noted that the performance obtained is a complex function of design and operating control, and the final design is a function of experience as well as theory.

In general, gaseous pollutants present in moderate concentrations, down to about 0.1% by volume, are handled effectively by liquid scrubbers as long as a suitable absorbing liquid is available. Scrubbing liquids include water for ammonia and hydrochloric acid gas, sulphuric acid for ammonia, methyl and ethyl amine solutions for hydrogen sulphide, sodium sulphite solutions and lime water slurries for sulphuric acid, and many others. The collected gases, in a number of the examples given above, are often stripped from the liquid phase with steam or by direct heating, and further treated, now as a more concentrated product, by a secondary process. The removal of hydrogen sulphide from natural gas and petroleum refinery products and its subsequent reduction to sulphur is an example of the use of scrubbing for concentrating the pollutant prior to conversion to a more useful or more readily disposable form.

Adsorption on Solids

Concentrations of pollutants can be far lower than 0.1%, possibly by several orders of magnitude as is the case for many odours. If the molecules are small and polar, then they can be adsorbed on solids such as silica gel, alumina or charcoal. Successful adsorbing solids like silica gel, although appearing to the unaided eye to be solid, actually consist of a multiplicity of minute pores, which greatly increase the surface area available. For example, the effective surface area of silica gel is of the order of several square metres for each gram of adsorbent. Solid adsorbents of this type can be re-used; silica gel, after saturation with water vapour, can be dried out (at 105°C) and re-used many times over*.

For large non-polar molecules such as the various organic compounds, including most odours, the most suitable adsorbent is 'activated' carbon. This is charcoal prepared by the destructive distillation of wood, coconut shell, briquettes or coal, which has then been treated (activated) with steam and sometimes inorganic chemicals. The total internal surface area of this material is much greater than silica gel, and can range from about 200 to 1200 m² g⁻¹ (square metres per gram). It can, therefore, act as an adsorbent for

---

*The blue colour in the laboratory grades of silica gel is due to the inclusion of small quantities of cobalt chloride ($CoCl_2$). When the silica is saturated with water, the cobalt chloride forms a complex with water, which is pink. On heating the water is driven off, and the cobalt complex returns to its blue form.

odours for periods of weeks or even months without replacement or regeneration. The way in which activated carbon acts is still incompletely understood. It can be regenerated by heating to 650°C in an inert (non-oxidizing) atmosphere for several hours, and subsequent reactivation.

Activated carbon can also be used for solvent recovery. The usual system is to use two adsorbers alternately and regenerate the solvent with steam. The important fundamental difference between a solid adsorber and a liquid absorber is that in a liquid absorption system the liquid is continuously removed and the surface regenerated as the liquid flows over the packing in a tower or other apparatus. In a solid adsorbent system, the solid itself has to be renewed when exhausted, and this is most effective in a batch system. Where continuous solids' renewal is used, there tends to be a deterioration of the solid adsorbent by attrition as this moves slowly through the apparatus.

Physical adsorption can be accompanied by reaction, in which case it is generally termed 'chemisorption'. Chemisorption is in general a more selective process than physical adsorption, with much stronger bonding between the gas and the surface of the solid, making regeneration of the adsorbent more difficult. In certain cases, the adsorbent serves as a site for reaction between the two adsorbed pollutants, or between an adsorbed pollutant and a chemical impregnated onto the adsorbent, rather like the colour changing gas detection systems discussed in Chapter 4. The solid adsorbent in these circumstances may play a catalytic role aiding the reaction, as many species are more reactive in the adsorbed than in the free state.

Catalytic decomposition on surfaces impregnated with a suitable catalyst is also used. Here the decomposition products may have no affinity for the adsorbent and are re-released into the gas stream. Such an adsorbent system can operate continuously and is only limited by the susceptibility of the catalyst to deactivation. An example is the breaking up of nitrogen oxides in automobile exhaust into nitrogen and oxygen, which are then re-released to the exhaust stream. The presence of lead from the 'anti-knock' added to motor spirits deactivates (poisons) such catalysts and these additives must not be used in conjunction with them.

Combustion

Combustion involves the treatment of combustible hydrocarbon air pollutants and carbon monoxide by their complete oxidation to carbon dioxide and water. In some cases, there is sufficient quantity of the hydrocarbon present to support combustion without the addition of further combustible material. In this case, the material is simply passed into a combustion chamber designed for the purpose (Fig. 5.3), and the gases are treated just like ordinary fuel gases, which are usually

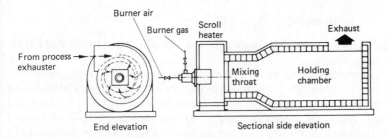

Fig. 5.3 Combustion chamber for control of combustible waste gases. (From Strauss,
W. *Industrial Gas Cleaning*. Pergamon Press (1966).)

burned by premixing with air (as in a Bunsen burner).

The chambers are designed to maintain the gases at 700°C for about
half a second, which time has been found in practice to give complete
combustion. If there is insufficient thermal energy in the hydrocarbons
in the gas stream, additional combustible material (either gas or light
fuel oil) is supplied. This can either be burned separately or, in the case
of gas, it can be mixed with the polluted waste gas.

In large works, where there are steam boilers, it is sometimes
possible to pass the waste gases containing the pollutant into the
combustion chamber of the boiler. However, unless the boiler has
been specially designed for this additional duty, severe problems can
arise. For example, if the waste gases contain organic sulphides, the
oxidation of these can give rise to large quantities of sulphur dioxide
and trioxide, and lead to corrosion and rapid deterioration of the
boiler. Waste gases containing appreciable quantities of water vapour,
which has a high specific heat, can reduce boiler temperature and also
affect boiler operation.

A notable example of a specially designed boiler is that designed for
burning black liquor in the paper pulp industry. These boilers also
burn the very noxious odours coming from the drying of the black
liquor, which is their fuel.

If there are surges in gas concentration, composition and quantity,
even combustion becomes difficult. Under such circumstances, which
are common in oil refineries, special 'flares' are used which have pilot
lights and supplementary fuel. As these flares burn considerable
quantities of aromatic (organic ring) compounds which produce a
smoky flame, there is a special provision for the injection of excess air
and steam, which reacts with the unburned carbon in the flame producing carbon monoxide and hydrogen:

$$C + H_2O \rightarrow CO + H_2$$
$$H_2 + \frac{1}{2}O_2 \rightarrow H_2O$$
$$CO + \frac{1}{2}O_2 \rightarrow CO_2$$

overall $C + O_2 \rightarrow CO_2$

In this way, smokeless combustion is achieved and a controlled flare is barely visible.

Combustion can be catalytically aided in circumstances where the emission does not contain significant amounts of catalyst poisons. The catalysts most commonly used are noble metals, e.g. platinum and palladium, or transition metal oxides such as cobalt chromium manganese deposited on a porous alumina support. The advantage of catalytic combustion is that hydrocarbons are oxidized at much lower temperatures (400 to 500°C) than combustion in a flame (700 to 800°C), and so will require less additional heat. This is particularly beneficial when the effluent contains insufficient hydrocarbon material to make the combustion self-supporting.

Catalytic oxidation is widely used for odour control from food processing, e.g. coffee roasting, and for the oxidation of organic vapours from paint baking and enamelling, both processes which have low emissions of potential catalyst poisons.

Control of Particulate Pollutants

The first widely recognized form of air pollution was smoke, fine carbon particles arising from incomplete combustion of fuels, and of inorganic ash arising from the non-combustible matter in the fuel. Even quite thick smoke plumes can represent relatively low concentrations of 2 or 3 g m^{-3}, which is of the order of parts per million by volume for the range of densities found in emissions. The clean air regulations in most countries require a reduction in the smoke concentration to 0.1 g m^{-3} or even less, of the order of parts per ten million. In the case of smoke, such concentrations are still visible to the naked eye.

In the most general terms, the control of particulate air pollutants involves passing the gas stream containing the particles through a chamber and permitting a force to act on the particles which takes them out of the gas stream. The method chosen depends on several factors including the nature of the plant operation; whether it is cyclic or continuous, and whether the emission is likely to vary at different times of the day. This is important as some dust collectors are more suitable for discontinuous operation than others, and some collectors are unsuitable for use with variable emissions. Information is also required about the nature of the particles, their size, shape, density, state (whether liquid or solid), chemical composition, and electrical conductivity. Finally, the properties of the gas in which the particles are suspended need to be known, particularly its temperature and chemical composition. For example, high temperature emissions can carry large amounts of condensible vapours in the gas phase. The most important of these would be water and acid vapours (e.g., hydrochloric acid, sulphuric acid and nitric acid), and their presence dictates the temperature at which the gas cleaning apparatus must be operated and, in some cases, the resistance to corrosion of the materials from which it is constructed.

It should be realized that the greater the efficiency of collection, the higher the cost both of the the collector and its operation. In general, this relationship is an exponential one, the cost increasing two-fold for an increase in efficiency from 90% to 99% and four-fold for an increase from 99.9% to 99.99%. The relationship of efficiency to cost is also reflected in the choice of method used, certain systems being both high efficiency and high cost without any built-in flexibility to reduce cost greatly by reducing efficiency. This is true of fabric filter systems, which would not therefore be chosen for use in a situation where the efficiency requirements were not high. Deciding what is the most effective system under a particular set of circumstances is a trade-off between the advantages of the clean emission and the cost disadvantage to the industry concerned. How these trade-offs are achieved will be considered in later chapters.

Gravity Separation

The simplest gas cleaning plant for particles is one which uses gravity to remove the particles from the gas stream. The device, called a 'Gravity Settling Chamber' (Fig. 5.4), permits particles to settle out of the gas stream into collecting hoppers, while the gases pass through the chamber at a reduced velocity. Solid or liquid particles suspended in the gas reach a terminal free falling velocity which is given by Stoke's Law for small droplets and particles, and is proportional to the product of the square of the particle diameter and the density difference between the particle and the carrier gas, and inversely proportional to the viscosity of the carrier gas. A 100 μm diameter particle of density 1000 kg m^{-3} will fall at a speed of 0.25 m s^{-1} and particles of 10 μm

Dust collecting hoppers

Fig. 5.4 Diagram of a gravity settling chamber for removing particulates from a gas stream. (From Strauss, W. *Industrial Gas Cleaning*, Pergamon Press (1966).)

diameter at 0.003 m s⁻¹. In practice, this means that the method is only useful for removing particles greater than 100 μm in diameter, smaller particles requiring the settling chamber to be impractically long. For smaller particles which, because of their ability to remain airborne, constitute the vast majority of problem particles, other methods must be used.

Electrostatic Precipitation
More effective precipitation can be gained by using electrostatic forces than by relying on gravity. A charged particle will move away from a similarly charged wire or plate towards one of opposite charge, usually earthed. This may be a tube or plate. Electrostatic precipitators are extensively used for particle collection on a very large scale; for cement dust from cement kilns, to some extent for collecting droplets of sulphuric acid mist in acid plants, almost universally for fly ash in power stations, and in most other situations where large volumes of gases are treated and high efficiencies have to be achieved.

The processes in an electrostatic precipitator are rather complex, and can best be described as follows. When a thin wire, central in an earthed tube (Fig. 5.5) or between two earthed plates, is charged negatively to a very high voltage (of the order of 40 to 60 kV), the air or

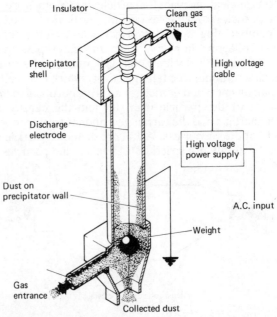

Fig. 5.5 Simple wire in tube electrostatic precipitator. (From Strauss, W. (ed.) *Air Pollution Control – Part I*, Wiley Interscience (1971).)

gas molecules immediately around the wire are bombarded by electrons released from the wire, and some molecules gain negative charges. In the electrostatic field between the wire and tube or plates, these negatively charged gas molecules (ions) move towards the positive (earthed) sections. If a dirty gas (containing particles) is passing through the tube or between the plates, the ions collect on the particle surface, charging the particles. These charged particles now move towards the earthed surfaces where they are discharged. After a time, a layer of the particles several millimetres thick collects. The collected layer or 'cake' is removed at intervals by tapping the tubes or plates, a process which is called 'rapping'.

Because of the very large scale of these units, and the very high voltages used, these processes are all carried out automatically. A typical plate type electrostatic precipitator is shown in Fig. 5.6. In actual practice, the processes are not as simple as in the above description. For example, the discharge of newly arrived particles at

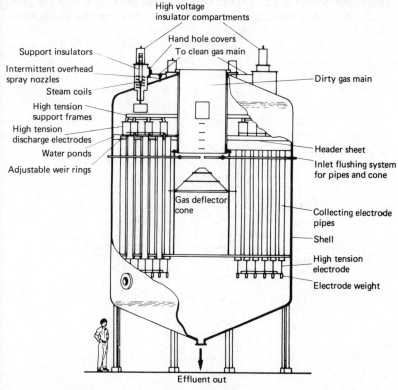

Fig. 5.6 Typical plate type electrostatic precipitator. (From Strauss, W. *Industrial Gas Cleaning,* Pergamon Press (1966).)

the plate, where there is already a deposited cake acting as a partial insulator, may be very slow. Indeed a charge can build up on the particle layer which is greater than the charge on the central wire, and 'back arcing' will occur, which strips some of the deposited particles and re-entrains them into the gas stream.

In spite of these problems, modern electrostatic precipitators operating under favourable conditions can achieve efficiences of 99% or better in the collection of fine particles, down to and below 0.01 μm. They can also be built to the efficiency required which allows for considerable cost saving.

Cyclones

A much simpler device, widely used on comparatively coarse dusts and powders such as sawdust from wood working, is the cyclone (Fig. 5.7). In the conventional cyclone, the dirty gases enter the cylindrical chamber to which a conical lower section is fitted tangentially. The gases spin downward and at the bottom of the cone, to which a hopper is attached, they reverse direction while still spinning and finally leave

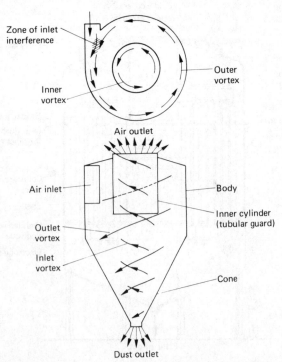

Fig. 5.7   Simple cyclone dust collector showing gas flow. (From Parker, A. *Industrial Air Pollution Handbook*, McGraw-Hill (UK) (1978).)

through the centrally placed exit pipe. The dirt particles are spun outwards and downwards by the movement of the outer layer of gases, finally being deposited in the hopper below. For industrial gas cleaning, cyclones are most useful for particles above 10 μm diameter; below this diameter the efficiency for a reasonable throughput falls away.

Fabric Filters

In the simplest terms, the fabric filter is a large scale version of the domestic vacuum cleaner. The effluent gas is caused to flow through the filter material and the particles are removed on this material. The mechanisms involved are more complex than direct sieving of particles out of the air stream. This is evidenced by the high efficiencies obtained for collecting particles which are smaller than the interstices in the filter fabric.

The basic unit of a fabric filter is the fibre and these are in general larger than the particles to be collected, and collection occurs as a result of the operation of several mechanisms (Fig. 5.8). Particles are

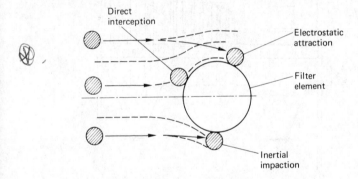

Fig. 5.8 Modes of particle removal by a filter element. (From Stern, A. C., Wohlers, H. C., Boubel, R. W. and Lowry, W. P. *Fundamentals of Air Pollution.* Academic Press (1973).)

directly intercepted by the fibre when the streamline containing the particle passes within one half particle diameter of the filter. Particles are impacted when the particle has sufficient inertia to stay on course when the gas streamline deflects around the particle. Particles in the smaller size ranges contact the filter fibre as a result of their own random motion (Brownian Motion) in the gas stream, and others are brought into contact as a result of electrostatic attraction. In due course, a layer of particles builds up on the surface of the fabric and a cake forms, which increases the efficiency of the filters but decreases the flow rate. The filter mat, therefore, has to be removed at intervals by shaking the fabric or reversing the airflow, or both.

A simple fabric filter, housed in a steel casing, is shown in Fig. 5.9. The fabric, which can be woven or felted, is usually arranged in tubes or flat sheets supported on wire frames. Groups of the tubes or flat sleeves are separated off into sections in which the cleaning process can be carried out separately. Thus, the fabric filter can continue operating while one section is being cleaned, the remaining sections now taking the whole of the gas flow. All large filter installations work completely automatically. Because of their simple basic structure, they need only occasional maintenance, replacement of torn tubes, and attention to mechanical faults.

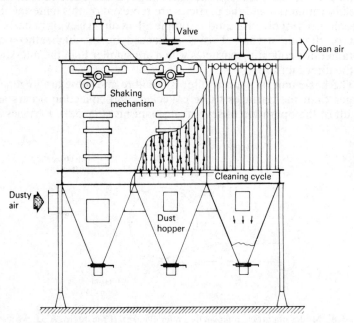

Fig. 5.9　Simple fabric filter baghouse. (From Stern, A. C. (ed.) *Air Pollution Vol. IV*, Academic Press (1976).)

Traditionally, fabric filters are called 'bag filters' or 'bag houses' because the earliest types over a century ago consisted of long bags, open at the top for entry of the dirty gases and tied at the bottom. At intervals a man entered the bag house, untied the bags, and collected the dust manually.

The temperature at which the fabric filters can operate is essentially a function of the characteristics of the fibres of which the tubes, or filter sleeves, are made. For commercial use at the highest temperatures glass fibre fabrics are available which will operate up to 270°C. Synthetic fibres such as Nylon*, Terylene†, Dacron*, Nomex* (a new

*Du Pont de Nemour registered trade name.　†ICI.

type of polyamide), or Teflon* (polytetrafluorethylene) operate at lower temperatures but give much higher filtration rates than glass. Gas emissions can be cooled by dilution to meet their requirements. Fabrics can be chosen to work for long periods, even under acid or alkaline conditions.

Wet Collectors
The action of rain in cleaning up the air is well known, and this can be used in a particle cleaning system which is usually called scrubbing. The simplest type of scrubber is the spray tower, where falling droplets collect the dirt particles. This is only effective for fairly large particles and so is often used as a pre-cleaner, particularly where an increase in humidity and cooling of the gases can help the subsequent cleaning process, as applies with electrostatic precipitators.

Packed towers, which were described for gas absorption, can also be used for particulate scrubbing, but are not usually very effective for the capture of medium-sized particles (0.5 to 10 μm diameter).

In dynamic gas scrubbers a film of water is sprayed on to a moving surface, such as the blades of a fan. In a centrifugal spray scrubber (Fig. 5.10), the gases are introduced tangentially into a cyclone in which droplets are sprayed outwards from centrally placed sprays. The droplets are spun outwards, being comparatively large, and intercept the dust particles. At the wall the wet film prevents re-entrainment of particles after capture.

A most effective type of scrubber is the 'venturi' scrubber, where the liquid is brought into contact (usually through sprays) with the dirty

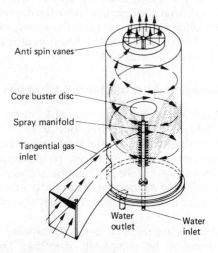

Anti spin vanes

Core buster disc

Spray manifold

Tangential gas inlet

Water outlet

Water inlet

Fig. 5.10 Simplified diagram of a centrifugal spray scrubber. (From Strauss, W. *Industrial Gas Cleaning.* Pergamon Press (1966).)

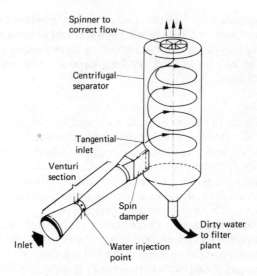

Fig. 5.11   High energy Venturi scrubber. (From Parker, A. *Industrial Air Pollution Handbook*, McGraw-Hill (UK) (1978).)

gas moving at high speeds through the throat of a venturi (Fig. 5.11). Venturi scrubbers require considerable energy to operate them because of the need to push the gases at high velocity through the system. It has been shown that to capture fine particles far more energy has to be expended than for coarse ones.

While scrubbers are widely used, they have serious disadvantages. One is that the liquid that has been used has to be disposed of or re-used, and in either case often requires extensive treatment, which is costly and may be difficult. The other is that the liquid itself is an expense even when, as in most cases, it is only water.

Disposal of Residual Air Pollutants – Chimney Stacks

It must be realized that no control system will remove all air pollutants from an industrial emission. Some fraction, 10%, 1% or even 0.1% of the pollutant material, remains in the waste gas stream and even this small proportion can, in some instances, be an appreciable absolute quantity, as in the example of the large sulphuric acid plant (Chapter 2), where the most effective methods of control with efficiencies well over 99.5% resulted in the emission of several tonnes of sulphur dioxide each day.

Residual pollutants are released to the atmosphere, ideally at levels that are considered to be completely harmless. This is frequently achieved by dilution using tall stacks. Tall stacks introduce pollutants into the atmosphere at sufficient elevation to ensure some dilution before ground level contact occurs. The effluent gases have a certain velocity and are generally warmer than the atmosphere. This kinetic energy and added buoyancy lift the emission above the top of the stack, where turbulence and diffusion mix it with the atmosphere. This dilutes the effluent and decreases the maximum concentration likely at ground level by an amount that is roughly proportional to the square of the effective height of the stack (where the effective height is the sum of the

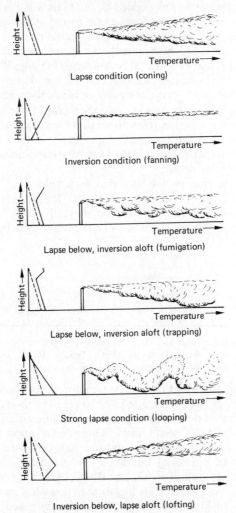

Fig. 5.12 Plume types associated with different lapse rates. (From Strauss, W. *Air Pollution Control – Part I*, Wiley Interscience (1971).)

actual height, the velocity rise, and the buoyancy lift of the plume).
Exactly what form the plume will take and how effectively it is diluted
before contacting ground is a function of the stability of the atmosphere
as measured by the lapse rate (Fig. 1.3). Fig. 5.12 shows the characteristic plumes corresponding to commonly observed lapse rate
conditions. At the theoretical (adiabatic) lapse rate, the atmosphere is
uniform and coning occurs. Under unstable conditions, when the lapse
rate is less than adiabatic, looping is observed. With a looping plume,
dispersion is greater but associated fluctuations will cause segments of
the plume to reach ground level close to the stack, giving elevated
concentrations for short periods. Under stable conditions, where the
temperature increases with height, the plume will have very little
tendency to disperse – the condition known as fanning. Where there are
discontinuities in lapse rate, the outcome for the plume will depend on
whether the discontinuity occurs above or below the stack height.

When the emission occurs above a thermally stable layer, then this
layer will serve as a barrier to its dispersion toward ground level, giving
a lofting plume. When the stable layer is above the stack, then the
plume is trapped beneath it and the two worst conditions, trapping and
fumigation, are the result.

When the terrain is not uniform, as it seldom is, other factors
interfere. The most common of these, for stacks, is increased
downward diffusion occurring on the leeward side of buildings. This
can affect the air layers, even at twice the building height (Fig. 5.13).

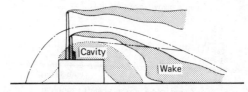

Fig. 5.13   Diagram to show the effect of a building on pollutants ejected at different
heights. (From Stern, A. C., Wohlers, H. C., Boubel, R. W. and Lowry, W. P.
Fundamentals of Air Pollution, Academic Press (1973).)

Consequently, the short chimneys common on residential and commercial buildings, which are adequate to get the chimney to 'draw' air
through the fire in the boiler or fireplace, are not sufficient to provide
good dispersal of the waste gases. The rule for good dispersal is to
make the chimney at least two and one half times the height of the
neighbouring or adjoining buildings. Unfortunately, for reasons of
cost and the aesthetic (architectural) appearance, commercial and
residential chimneys are rarely higher than just a few metres above the
highest point of buildings. As we have already noted, tall stacks in
solving a local problem often create a new problem somewhere else, as
evidenced by acid rains in North West Europe and North East

America.

The choice of control measures on any particular plant will depend on what regulatory requirements are to be met, and what the most efficient and economical way of meeting these is, taking due consideration of the concentrations and composition of the effluent to be cleaned. Similar industries using similar raw materials have common control problems. The current solutions of some of these problems for particular industries will now be considered.

Applications to Control Technology

Roasting of non-ferrous metal ores

In Chapter 2 it was pointed out that the non-ferrous metals – copper, lead and zinc – are obtained from sulphide ores which are all initially treated by roasting the ores, producing large quantities of sulphur dioxide. Subsequent treatments also give rise to sulphur dioxide. If the sulphur dioxide concentration is greater than 3.5% by volume and is produced continuously, it can easily be oxidized by catalytic oxidation, producing the trioxide ($SO_2 + \frac{1}{2}O_2 = SO_3$), which is then absorbed in dilute sulphuric acid to give the more concentrated acid which can be used commercially. However, the gas from the smelter generally contains other impurities including several such as arsenic halides, sulphides and oxides, mercury and selenium, which can poison the catalyst and must be removed before conversion.

If the sulphur dioxide concentration is high (>10%) and oxygen concentration is low, then direct reduction can be used to give elemental sulphur, a solid end product that is comparatively easy to handle. Like the oxidation process, direct reduction requires a continuous flow through – a condition which is not met in most copper smelters which are batch processes. When these requirements are not met, then pre-treatment processes using absorption to concentrate the SO_2 are necessary, greatly increasing the cost of the procedure.

The choice between oxidation and reduction treatment routes will depend on several factors including the market for the product. Sulphuric acid has a use as fertilizer feed stock, and sales of this secondary product can offset the cost of treatment provided the distance to markets is not too great. If no market for the acid can be found, then it will have to be neutralized, making the oxidation process overall more expensive than reduction to elemental sulphur, which can be stored safely without pre-treatment. Another factor which may influence the decision is the allowable sulphur content of the final emission. Sulphur emissions are significantly higher in a reduction plant and may require additional further treatment.

Waste gases from thermal electricity generating stations
Much of the world's electricity is produced from the combustion of
coal, and the waste gases contain fine particulates (fly ash), oxides of
nitrogen and sulphur dioxide. Oil fired stations also produce oxides of
nitrogen, sulphur dioxide and some particulates. Particulates from
coal firing originate from the non-combustible mineral material in the
coal; oil contains much less mineral material and therefore has a
significantly lower particulate emission. Sulphur dioxide in both cases
comes directly from the oxidation of sulphur in the fuel – coals contain
from 0.5 to over 5% sulphur and oils from 0.5 to 4%. A 1% sulphur
coal will, on combustion with 15% excess air, give 0.06% sulphur in
the effluent gas, whereas a 4% sulphur coal will give 0.25%. Fuel oil
will give slightly lower values for the same percentage sulphur in the
fuel. The obvious answer would seem to be to concentrate on low
sulphur fuels, but unfortunately these are in short supply.

In the absence of such low sulphur feedstock, the possibilities are
treating the fuel prior to combustion, or removing the sulphur
afterwards. The first possibility seems attractive because the concentration of sulphur is much higher in the fuel than in the gas
emission. However, desulphurization of fuels has proved difficult and
to date more success has been obtained using gas cleaning techniques.
The alternative of using high stacks, although locally effective, has
become increasingly unacceptable, as we have already seen.

Many processes have been suggested for the removal of sulphur
oxides from stack gases. The earliest which was successful involved
scrubbing the flue gases with the slightly alkaline waters of the Thames
River in London. While this proved feasible for two medium-sized
power stations in the 1930's, extension of the scheme was not possible
because the increased sulphate content of the river affected marine life
and made the water unsuitable for industrial use downstream.
Numerous other processes for scrubbing flue gases have been
suggested, including absorption in calcium carbonate, calcium oxide,
magnesium oxide, or mixed magnesium/manganese oxide slurries,
absorption in sodium sulphite solution, adsorption on dry limestone,
and catalytic oxidation.

Most interest at present in the USA, where emission standards for
SO_2 are tightest, is in absorption of SO_2 in wet lime (CaO) or limestone
($CaCO_3$) slurries which are then discarded. One of the main problems
of this method has been disposal of the waste sludge produced – a
mixture of sulphates, sulphites and limestone – which has poor settling
characteristics and has proved very difficult to deal with. The
magnesium oxide process avoids the waste problem by regenerating
the magnesium salts after SO_2 absorption and either oxidizing or
reducing the now concentrated SO_2 regenerated. This process requires
several additional steps to the limestone absorption and discard

process; not only does the SO_2 have to be treated twice but for coal fired boilers particulates must be removed before the gas treatment commences.

Another possibility being considered is catalytic oxidation, but here the main difficulty is that the sulphur dioxide concentrations are so low (0.2% compared with 7% for normal catalytic conversion plants as used in the non-ferrous metal industry). Catalysis requires high temperatures, around 450°C, and the stack gases must be free from particulates. This means particulate removal at high temperatures plus, where necessary, reheating of the gas prior to its entry into the catalysis bed. None of these systems have proved to be completely satisfactory for the removal of SO_2 from power generating plant emissions.

Nitrogen oxides are produced from oxidation of atmospheric nitrogen at the high temperatures in generating plant furnaces, and by combustion and release of nitrogeneous components in the fuel. Control of oxides of nitrogen from large boilers is a comparatively new problem, which has arisen because of the recognition of the contribution these make to photochemical smog. There are at present no processes, even at the experimental stage, which could be used for taking nitrogen oxides out of flue gases, and the approach has been to reduce NO_x emissions by modifying the combustion. The reaction for nitric oxide formation is:

$$N_2 + O_2 \rightarrow 2NO$$

The extent to which this reaction occurs in a boiler is a function of the temperature of combustion, which depends on the boiler design and the fuel used, and on the amount of excess air present. Combustion of coal in conventional boilers will give 360 grams of nitrogen oxides per million BTU's at 3% excess air, corresponding to an emission gas concentration of about 570 ppm. Many existing boilers give values much higher than this, up to 1200 ppm. The corresponding values for fuel oil are 310 grams per million BTU's with a concentration of NO_x in the emission gases of 530 ppm.

As the furnace temperature increases, the extent of reaction between N_2 and O_2 also increases; methods which can keep the temperature of combustion down without loss of efficiency in the boiler are, therefore, desirable. Such methods include tangential firing, where the furnace flames are positioned so they do not impinge directly on one another; this gives a more uniform temperature throughout the combustion chamber. The other approach is to reduce the combustion air supply so that there is just sufficient oxygen to burn the carbon but no excess is available to react with the nitrogen.

Particulates are a major problem in coal fired boilers. The ash content of coal varies from 20 to 40% and when such coal is ground prior to combustion the gaseous emission can contain up to 80% of the original ash content as fly ash. The most suitable method for removing this fly ash is electrostatic precipitation. Efficiency has been found to be a function of the sulphur trioxide (SO_3) concentration in the gas, which is a function of the sulphur content in the coal. Higher efficiencies are obtained for higher sulphur coals, and in some circumstances it may be necessary to add SO_3 or sulphate to the stack gas.

Automobiles

The conventional automobile, driven by an internal combustion engine, has presented a special problem in air pollution control. The proposals for controlling the different air pollutants use many of the processes that have been discussed, in control units suitable for attachment to a vehicle. It was pointed out in Chapter 2 that air impurities from a conventional automobile are unburnt hydrocarbons, carbon monoxide, nitrogen oxides and particulate matter. All are emitted through the exhaust, but the hydrocarbons are also emitted through other parts of the engine. The approximate distribution of hydrocarbon emissions in uncontrolled cars is 60% through the tail pipe, 20% through the crank case, and the remaining 20% about equally from the carburettor and the fuel tank. These percentages will vary – fuel tank evaporation will be low from all cars in cold weather, and older cars with worn cylinders and piston rings will have high crank case emissions. Crank case and fuel tank emissions were the first to be controlled. Crank case emissions, which have simply been bled from the crank case to the atmosphere, are now controlled by passing the crank case gases back into the system, either into the air cleaner or into the intake manifold, or both. Evaporative emissions from the fuel tank and from the carburettor reservoir are controlled by venting the fuel tank and carburettor reservoir through a bed of activated carbon. When the motor spirit from the stationary vehicle evaporates, for example when the vehicle is standing in the sun, it is absorbed in the bed of activated carbon. On restarting the automobile, air (after being passed through the air cleaner) is drawn through the activated carbon where it picks up the motor spirit and returns it to the engine intake, giving a slightly richer mixture in the combustion chamber. Other carburettor losses are controlled by eliminating conventional vents to the atmosphere, and leading all vapours into the inlet air stream or to an activated carbon canister.

The exhaust gas emissions present a more complex problem. Essentially, better combustion will reduce the unburnt hydrocarbons and carbon monoxide, but will not change the particulate matter and could increase the oxides of nitrogen because of the higher combustion temperature. Modifying the fuel and increasing its volatility will help, and so LPG (liquified petroleum gas) has been introduced for some vehicles, such as taxi cabs and local delivery vehicles, used mainly around cities close to sources of LPG. However, the limited mileage and space and weight of the LPG cylinder (as well as the limited total supply of LPG) have restricted its use.

The removal of lead compound anti-knock agents will reduce the lead contaminated particulate emissions. On the other hand, the use of leaded fuels helps engine operation and enables a much greater range of motor spirit to be brought up to the high 'octane' (anti-knock) characteristics required for modern, high efficiency, high compression engines. Using engines with lower compression ratios would help in reducing the necessary lead additives, but such engines are also less efficient and require more fuel for the same mileage.

Better combustion chamber design and greater control of the air/ fuel ratio, so that leaner (optimum) fuel mixtures are used during all driving modes, has gone a long way to meeting the requirements for control of hydrocarbons and carbon monoxide. This was sufficient to meet regulatory requirements for vehicles in the USA until the 1974 model year. The 1975 requirements, however, have necessitated the use of exhaust treatment devices on some vehicles (Table 5.1).

Table 5.1 Some sample emission limits adopted for California and the United States (g per m)

	Uncontrolled* Vehicle	1970† California	1974* USA	1975* USA	1981* USA
Hydrocarbons	5.9	1.4	2.1	0.9	0.25
Carbon Monoxide	60.0	14.0	24.0	9.3	2.1
Nitrogen Oxides	2.2	no std.	1.9	1.9	0.6

(Source: Stern, A.C., Ed. *Air Pollution, Vol. V*, Chapter 12 (1977). US EPA *A compilation of Air Pollution Emission Factors* (1976).)

† California Cycle – a seven mode test cycle including idling, cruising at two speeds, acceleration and deceleration.
* Cycle (LA 4 cycle) – a simulated trip of 12.1 km taking 23 minutes and representative of the early morning peak hour driving pattern in Los Angeles.

These standards are still being debated and the mechanism exists for waiving the CO standards and also, for a percentage (5%) of a manufacturer's cars, the NO_x standard.

Catalytic oxidation using platinum and palladium catalysts has been successfully used for conversion of exhaust hydrocarbons and carbon monoxide into carbon dioxide and water. Because of problems of catalyst poisoning, the fuel used must be lead free or the lead particulate removed before the exhaust gases pass through the catalyst bed. This latter problem is avoided in the direct combustion system used on some engines, in effect continuing combustion in the exhaust by introducing extra air and maintaining the temperature.

Control of nitrogen oxides has proved to be more difficult. Many of the measures taken to control hydrocarbons and carbon monoxide lead to increased nitrogen oxide emissions. The approach to NO_x control used in other combustion systems, of altering combustion characteristics, has proved with the internal combustion engine to be more expensive in fuel terms.

Some reduction of the nitrogen oxides is achieved by partial recirculation of the exhaust gases, but the more stringent requirements need some form of catalytic or thermal reduction of the nitrogen oxides. The technical difficulties of meeting the required level were one of the reasons why the introduction of the stringent NO_x standard promulgated for 1976 model cars had to be postponed (Table 5.1). It may not be possible to meet the more stringent regulation without the use of reduction catalysts to convert NO_x to nitrogen. Reduction catalysts require a fuel rich mixture in the exhaust and therefore impose a fuel penalty. This penalty can be minimized by a precise control of air/fuel ratios, at the same time increasing the complexity of the system and its capital cost.

There are no emission standards in force for particulates from cars, or for the concentrations of the element of concern in these particulates – lead. However, more stringent hydrocarbon and NO_x standards requiring the use of catalysts necessitate the elimination of lead, either from the fuel or by filtration from the exhaust stream. In addition, regulations have been passed in most countries to limit the amount of lead that can be added to automobile fuels; such regulations have the effect of limiting total lead emissions.

Alternative Engines
Instead of dealing with the exhaust emissions formed in a conventional spark ignition (Otto) engine, it is possible to use alternative engine types. The most important of these, which was also discussed in Chapter 2, and is widely used for trucks and buses, is the diesel engine. Diesel engines for private automobiles are also available commercially, but for various reasons have not proved popular although in some countries, for example Germany, they are almost universally used for taxis. Diesels have the advantage not only of producing less carbon monoxide, hydrocarbons, nitrogen oxides and, under certain circumstances, particulates per kilometre but also of using less fuel (25% less for same vehicle weight) and having a longer engine life. Emission controls are still required on diesels if they are to meet the new NO_x standards, and the technical difficulties in implementing

these controls have resulted in a waiver for the NO_x standard on model years 1981-4. Methods suitable for the internal combustion engine are not as effective with the diesel – reduction catalysts will not be effective as they require a fuel rich exhaust, and quenching by exhaust gas recycling is not as effective because of the difference in the combustion system. Changes in the combustion system considered to date reduce NO_x emission at the expense of increased hydrocarbons and carbon monoxide.

Although particulate emissions are not significantly greater from diesel as compared with gasoline engines, their nuisance value is much higher. This is a result of the predominance of soot in the emission and also its unacceptable odour. To date little success has been achieved using catalytic oxidation, and the only means of controlling both odour and soot have been careful control of combustion to avoid fuel rich zones, modified fuel injection, adding barium compounds to the fuel, humidification of inlet air, and as an extreme measure, always working below peak power.

A development of the spark ignition engine which gives much better combustion is the 'stratified charge' engine. Here, a small spherical chamber with a volume of about 6 cm³ is placed next to, and is open to, the main combustion chamber of the cylinder. A very weak mixture of air and fuel (20:1 compared with 15:1 or less for conventional engines) is introduced into the main chamber and at the same time, using a separate carburettor, a very rich mixture (about 4:1) is introduced into the small spherical chamber and flows through the connecting opening into the main chamber, where it forms a rich cloud with the lean mixture. On ignition, this rich mixture burns immediately and the flame spreads through the lean mixture, giving excellent overall combustion, lower maximum temperatures, and reduction in air pollutants. A commercial development of this principle has produced, in operation, 1.23 g km⁻¹ carbon monoxide, 0.13 g km⁻¹ hydrocarbons and 0.50 g km⁻¹ nitrogen oxides. Even after 80 000 km, the emissions were only 1.60 g km⁻¹ carbon monoxide, 0.16 g km⁻¹ hydrocarbons and 0.60 g km⁻¹ oxides of nitrogen, which is well below the 1975 United States requirements for all the pollutants and approaches the ultimate limits suggested for the oxides of nitrogen. The fuel economy is, unfortunately, not as good as for the conventional engine.

6 Non-Technical Aspects of Control

Previous chapters have provided us with evidence as to why air pollution control is desirable in certain circumstances. We have also seen that we possess much of the technological expertise required to carry out that control. Neither of these factors is sufficient in itself to ensure pollution control will occur. What is necessary in addition is a political decision on what level of control is desirable, which balances the benefits of control against costs. Having made a decision on what to control, and to what extent, we then need to implement it by codifying it in law and setting up an administrative procedure to ensure the law is administered and enforced.

This all sounds very simple and straightforward but, as already discussed in Chapter 3, the relationship between effects and pollutant levels is by no means a simple one, and our level of knowledge, especially regarding chronic and synergistic effects, is far from complete. In the light of the quality of information presently available, chosen control levels should be flexible and open to re-determination as new information becomes available. The uncertainty of air pollution criteria also means that the direction of any reassessment of control measures should not necessarily be in the direction of stricter controls.

Currently, most countries have taken some air pollution control measures, generally in response to an air quality situation which has been judged to be unsatisfactory. Unsatisfactory conditions covered by such measures include both the concentration levels that produce acute episodes and those that are considered to be detrimental in the long term. Such measures explicitly or implicitly include an estimate of what level of risk to the population or environment is acceptable. This acceptable level will depend on a complex array of social, political and economic, as well as scientific, factors and will therefore vary from country to country. For example, in the developing world where life expectancies are low as a result of poor nutrition, health care and housing, controlling air pollution at the expense of growth will have little or no priority, and would have little obvious benefit. In the developed countries, the question that arises is who should be protected and to what extent. For instance, should the pollution level in the most highly industrialized areas be low enough to protect the weakest members of society, if this is possible, or should such people be recommended to live in areas with low pollution levels? Further, should society protect the self-polluting smoker whose level of intake of many major pollutants is higher than that receivable from the worst city atmospheres, and finally, are minor reversible effects to be considered adverse when they produce no long term deterioration?

The climate of an area will also influence the level of control considered necessary. The likelihood of photochemical smog, for instance, is much higher in regions with high insolation. Areas with lower sunshine hours may not consider it so necessary to control contributing pollutant emissions. Similarly, areas with a high degree of air turbulence do not necessarily require the same strict levels of control as regions with calmer wind conditions or those areas subject to frequent temperature inversions.

Air Pollution Management

In any air pollution control programme the first problem is to establish the criteria on which the control decisions are to be based. We have touched on some of the difficulties involved in doing so previously, particularly when it comes to health effects on human subjects (Chapter 3).

For living systems that can readily be experimented with, such as plants and some animals, it is possible to state what level of what pollutants will have specific effects. The relationship between level and effect can only be stated in statistical terms because the resistance to damage of individual specimens of the same species differs widely. The same statistical relationship will hold for effects on humans. However, in this case, it is not possible to assign a numerical value because this would require the exposure of human subjects to detrimental levels of pollutants.

The thing we are most interested in, as far as human health is concerned, is the relationship between exposure and health effects in a real situation such as in an urban airshed. This information is virtually unavailable at present for several reasons, the most important of which are outlined below.

1 Exposure levels available are not representative of real exposure as these are obtained, in most cases, from a limited number of fixed samplers, whereas our subject is mobile.
2 Exposure in a real world is difficult to define. Are we, for instance, looking for correlations between effects and the frequency that a particular pollutant exceeded a pre-determined value, or between effects and pollutant concentrations averaged over a pre-determined time? How do we separate pollution induced effects from socio-economic factors such as living conditions, diet, race and social habits?

3   It is possible that populations exposed to pollution over a period of time may have adapted and, as a result of exposure, have a greater resistance to elevated levels of pollutants, posing a problem not only for the effects relationship but also raising the question of whether the ability of human populations to adapt to changing conditions should be allowed for in setting acceptable levels or not.

4   Urban air contains a complex mix of pollutant gases which may act alone or additively or synergistically. The relating of a pollutant concentration or mix of pollutant concentrations to a particular effect is greatly complicated by ignorance about synergism, and some doubt might be cast on earlier conclusions regarding cause and effect.

To illustrate the last point, in the majority of studies on sulphur dioxide related effects, only sulphur dioxide and particulates were measured, and in some cases a third related factor was taken as a measure of their concentration. No certainty can be attached to the conclusions of these studies. The effects observed may well be related to another factor varying in the same way which was not measured, or to some other pollutant acting alone or synergistically with SO_2 or particulates. Fig. 6.1 illustrates very clearly how an obvious conclusion based on a good correlation between pollutant and effect can be wrong. The evidence prior to the Second World War strongly implicates pollution as the cause of respiratory mortality, whereas the fall-off in pollution during the War unaccompanied by any change in mortality indicates that the cause was elsewhere.

Air quality goals or standards are therefore, of necessity, based on

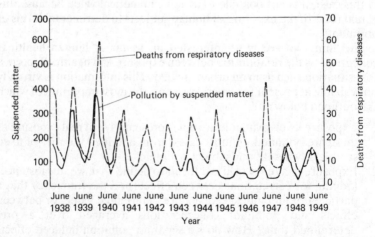

Fig 6.1   Graph of air pollution levels vs. death rate for Dublin. (From Leonard, A.G., Crowley, D. and Belton 'Atmospheric Pollution in Dublin during the years 1944-1950', *Royal Dublin Soc. Sci. Proc.* **25**, 166 (1950).

rather uncertain criteria and for this reason alone should be considered as flexible and open to change as new data become available. The goals themselves, like the criteria, need to be expressed in a statistical form because atmospheric pollutant concentrations are subject to many unknowns, including the uncontrolled vagaries of the weather, which make it impossible to set a concentration of pollutants which shall not be exceeded. The usual method of covering this is to set the goals as levels that should not be exceeded more than a certain number of times in an average year. This still leaves the possibility of exceptional conditions producing a higher frequency. A more rigid standard would have to include within its means of enforcement the possibilities of plant shut-down and/or reduction in industrial activity when concentrations exceed or are predicted to exceed certain levels. The concentrations chosen as cut-off values are averaged over a period relevant to the biological lifetime of the pollutant.

Setting air quality standards or goals involves a determination of what levels of effects are acceptable. This might be quite straightforward when the potential damage is to cash crops, whose value can be directly compared to the cost of control. It is a much more complex decision when the effect involves things that cannot readily be costed, such as a health disadvantage to a small sector of a community or an aesthetic deterioration, e.g. decrease in visibility. The cost of achieving a particular quality goal can always be estimated. The benefit of doing so, however, cannot be so easily quantified. This is in part why the problem arose in the first place in that things adversely affected by air pollution have no monetary value in the market place, and in particular no monetary value to the polluter. Society must therefore make a decision, trading off effects against the cost of control, and set up the mechanisms to meet these levels of pollution in the most cost effective way. This not only involves decisions on what level of effects is acceptable in the light of the cost of implementation, but what safety margins should be included in the standards in each case. Having established the level of air quality desired, it is necessary to determine the means of achieving this, leaving a suitable margin for future growth and development. To be able to look at possible strategies rationally, we need to have a thorough understanding of the airshed of interest. This means we need to know how the meteorology of the airshed interacts with the emissions into it to produce the concentrations of pollutants observed. This effectively means establishing some sort of model for the airshed so that the overall reductions in emissions necessary to produce the required atmospheric concentrations can be determined. In the first instance, a proportional reduction in emissions might seem to be logically connected with an equivalent reduction in concentration. However, as we have already seen when considering photochemical smog, the interactions between pollutants are sufficiently complex to invalidate such a simple approach. In fact, no universal model exists that is applicable to all airsheds and even models developed for and validated in particular airsheds are not without their problems.

Once the desired degree of reduction of pollutants in an airshed is determined, it remains to find the most cost effective strategy for achieving that reduction. The cost must take account not only of the costs of emission control equipment but also of the costs of ensuring compliance.

Several control strategies can be identified. One is to require the use of existing technology in present and new plant to produce the minimum possible emission to the air, termed 'best available technology'. Such an approach can be modified to take account of cost factors or special difficulties experienced by small or older plants, and is then generally referred to as the 'best practicable means'. This approach generally requires a prescribed method to be used to achieve a prescribed level of emission, based on a knowledge of available technology and the emission levels obtainable with this technology in different industrial plant. The overall result is not necessarily atmospheric concentrations that are acceptable in terms of effects, nor is account taken in these methods of making the best use of the assimilative capacity of the atmosphere. Best available and best practicable means are not, therefore, the most cost effective methods of control. They are, on the other hand, much easier to enforce than the airshed management approach discussed below.

Airshed management methods are based on managing emissions into an airshed to meet pre-determined air quality standards. This can be done by imposing emission controls on pollution sources in the airshed to enable the standards to be met under the most unfavourable meteorological conditions, or by having a sliding or variable scale of control allowing emissions to be higher when meteorological conditions ensure low pollution potential. In all cases, the decision on the level of emission control required involves reference to some model of the airshed that relates concentrations to emissions under the local meteorological conditions. How the burden of emission reduction is spread among individual emitters will depend on cost factors. It is, for instance, often more cost effective to put more severe controls on large emitters and not control small sources at all. On the other hand, it may be considered more equitable to share the load by cutting down relatively less on many sources. Other possibilities involve specifying fuel characteristics allowable for certain processes and/or specifying stack heights although, as we have already seen, this latter step may transport the problem elsewhere. When the required reduction is greater than can readily be achieved by existing technology, a time limit for meeting the requirements can be set, leading to a forcing of new technology or, failing this, a closing down of plant. One of the main criticisms of the best practicable means approach has been that it contains no incentives for technological improvements, in contrast to an air management approach.

A certain amount of flexibility can be introduced into an air management scheme using variable emission scales to optimize the use made of the dispersive capacity of the air. As the highest concentrations only occur under the most adverse weather conditions, it is not necessary to ensure emissions are always below the level that would produce these concentrations. When adverse weather conditions conducive to high pollutant levels are predicted, or when airshed concentrations indicate that allowable levels of pollutants will be exceeded, an alert can be declared and more severe levels of emission control can be applied. This could be done, for example, by requiring partial shut-down on the part of the largest emitters, or the temporary switch over to a less polluting fuel during an alert period. On the individual level, variable control might involve restrictions on driving in certain areas or bans on household or garden incinerators on particular days or at certain times of the year.

In most countries, some combination of the approaches discussed is used. Where air quality standards have been imposed, best practicable means approaches are frequently applied to source emissions. This both allows for ease of enforcement and provides a uniform base level of control so that cleaner areas do not provide advantages to industry that more polluted areas cannot offer. It also ensures that a minimum degradation in air quality occurs with development and growth by placing emission controls on all new industrial plant. For highly polluted areas where the base level of emissions control will not bring air quality within the standards, or where new developments will cause these standards to be exceeded, then other measures have to be taken. Such measures could take the form of stricter controls under adverse conditions, or require a higher level of technological control, or ultimately require shut-down of older plant to meet the standards. For an airshed which is just meeting its air quality standards, new development will be only possible with new technological development or if older industries are shut down, leading to the possibility of existing companies being bought out for the sake of their emission rights.

Some countries such as the UK have no air quality standards for any pollutants but operate on emission controls only. These controls are much stricter in areas which have high pollution levels, such as London and Coventry, and can be considered to be designed to produce air of a required quality, even though this quality has not been specifically defined. Emission controls without associated air quality standards are also used in most places as the sole means of controlling the less common toxic emissions associated with specific industries.

A question arises as to whether air quality requirements should be uniform within national boundaries, or whether there are areas that should be specially protected. The principle of having zones of special air quality where much higher air standards apply has been accepted by the USA for areas of natural beauty or high aesthetic value, such as national parks. Industry located near these areas is subject to much more stringent standards then elsewhere to ensure effectively that no degradation occurs.

An intermediate situation exists in areas which already have good air quality but are looking to expand their industrial base. The choice exists of giving industry in these areas an economic edge by having very lax restrictions on emissions, or taking measures to maintain existing air quality as far as possible. There are factors that make the latter approach a reasonable one. In particular, it is much cheaper to incorporate emission controls into new plant than to add them on to old, so that the overall cost of control in the long term is less for the new development. There are also the arguments regarding unfair commercial advantage mentioned earlier.

Apart from achieving levels of pollutant concentrations necessary to meet the standards required to protect health and property to an agreed level, there are also other measures that can be taken to minimize the impact of pollution on surrounding areas. For example, the impact of industry on residential areas can be minimized by careful planning. Industry can be located in an industrial estate with a buffer zone of vegetation between it and adjoining residential areas. The location of the estate can be chosen so that maximum use is made of the local meteorology to disperse pollutants. It must, however, be remembered that the main contributor to the level of pollution in our cities is the individual citizen and his motor car. We will return to land use planning as a factor in control later in this chapter.

All of these approaches are applicable and have been used successfully on the local and national level to mitigate the effects of pollutant emissions. On the international level, the solution of pollution problems is more complex. The international problems of particular concern are those that involve pollution damage in one nation as a result of emissions from another. The most critical problem in this regard is the transport of sulphur dioxide from North Eastern America to North Eastern Canada, and from the highly industrialized countries of Western Europe to Scandinavia. Considerable effort in both the European and North American situations has been put into joint studies of the transport and dispersion of sulphur dioxide, but no negotiated agreements exist between any of the countries involved regarding emission control. The most significant step has been a Memorandum of Intent between the USA and Canada foreshadowing a negotiated air quality agreement. Nothing of this kind has occurred in Western Europe.

On the global scale, there seems little hope of a world-wide consensus being reached on carbon dioxide emissions, although things are more hopeful for the restriction of chlorofluoromethanes if it is universally accepted that these are a potential danger. The essential nature of the carbon dioxide producing industries, particularly in the energy area, and the technical difficulties and cost of controlling CO_2 are the main reasons for considering a reduction in output to be unlikely.

Implementation of Control Programmes

Once the programme for pollution control has been decided, then a means of ensuring its implementation is required. Satisfactory implementation has two essential features: first, there must be suitable laws to ensure enforcement is possible, and secondly, there must be a management body that polices compliance with those laws. In addition, it is necessary to determine the efficiency of the strategy chosen in achieving the air pollution abatement desired, and to have procedures available whereby changes can be made in the level of control required if this is needed. Without specific legislation for the control of air pollution, the only recourse in law is to sue for compensation for actual damage done. Legal procedures for damage have proved too ad hoc to lead to an improvement in air quality and therefore have not proved suitable for implementing the strategies discussed.

Laws aimed at controlling air pollution fall into two categories which reflect the two approaches to control strategies. The first category specifies the emissions requirements on all major air pollution producing activities. The specification could include, as well as the allowable emissions from a process, the fuel type, the chimney height and the type of pollution control equipment to be used. This method defines best practicable or available means in law and, as a result, is easy to enforce but lacking in flexibility and particularly unresponsive to changing requirements.

In the second method, the law is used broadly to specify the outcome required, for example, an air pollution standard or, in even more general terms, the level of effects aimed for, thereby making allowance for the unknowns in the criteria. Best practicable means can also be covered by broad laws of this second type, the actual means to be used not being defined in the law. In countries where this more general approach is used, the necessary emissions control is obtained by giving the management authority regulatory powers in the legislation. Not infrequently some mix of the two methods is found, with emission

controls being enshrined in law for selected sources such as automobiles and statutory powers being given to an authority to control the
common emissions from stationary sources. Examples can be found in
the legislation of the countries of the European Common Market, the
USA, Australia and Japan. Emissions legislation may also be used to
set limits on the emissions of unusual toxic pollutants that are not
universally found in all airsheds.

The earliest regulations for pollution control were applied to cities
where adverse effects had been observed. Examples are the declaration of smokeless zones in Coventry in 1951 and London in 1955, when
restrictions were put on the type of fuel allowed. This brings us to the
question of the area of operation of regulations and regulatory
authorities. Should air pollution be a national responsibility with
regulations applying nationwide, or should it, at the other end of the
scale, be a local responsibility. In practice, various levels of
government are involved in control of air pollution and, in some cases,
in legislation as well, either as a result of powers vested in them or
delegated to them. This can lead to differences in the stringency of
control within the same national boundaries, with national and federal
legislation being supplemented at the state or provincial level and
again at the local level.

The way different governments implement their control strategies is
best seen by considering a few examples. For instance, in the UK the
Clean Air Acts operate on the best practicable means principle, taken
as meaning, in the words of the 1956 Clean Air Act, what is 'reasonably
practicable having regard among other things to local conditions and
circumstances, to the financial implications and to the current state of
technical knowledge'. Detailed emission standards are not specified in
the relevant Acts. The job of administration and regulation is
undertaken for some 61 listed processes on the national level by the
Alkali and Clean Air Inspectorate; all other clean air regulations are
administered at the local level. Local government responsibilities
under the Acts include: the control of smoke, dust and grit, the control
of chimney heights, and the declaration and control of smokeless
zones. The UK has no air quality standards legislation.

In the USA on the other hand, a combination of air quality and
emissions legislation is employed at several levels of government.
Since 1967 pollution control has moved from being purely a State
matter to come increasingly under Federal jurisdiction. Air quality
standards are set under Federal Clean Air legislation for the six major
pollutants and serve as the minimum standard the States are expected
to meet. The States then have the freedom and the responsibility to
determine how these standards will be achieved, with the requirement
that Federal approval be obtained of their implementation plans. The
States have the power to delegate authority further to the local level.

Emission standards are also covered by Federal legislation in certain areas. For example, Federal legislation covers emission standards on automobiles and hazardous pollutants, with the States having the power to introduce more stringent requirements as has been done for automobile emissions in California.

In Australia, air pollution control has been and still is a State matter, the only Federal legislation to date being that related to automobile emissions. Existing control legislation in Australia is of the best practicable means type with the exception of the State of Victoria, which has recently (1981) passed air quality standards legislation coupled with emission control requirements for scheduled premises.

A body of legislation related to air pollution does not necessarily mean that effective or efficient control will be attained. There are several other factors involved, not least of which is the power and independence of the regulatory body, the level of its funding, the quality and enthusiasm of its staff, and the degree of public support it receives.

The function of the regulatory authority will vary as a function of the law it is operating under. In general, its function will be to police the regulations, which it may or may not have been instrumental in setting up, and to determine the effectiveness of these same regulations in achieving the desired objective. In addition, the regulatory authority, or some other body set up for the purpose, will be responsible for planning functions including the procurement and assessment of new criteria data, the assessment of the effect of future developments on air quality, and the long term implications of changing energy use patterns (Chapter 7).

The policing function of a control authority is to ensure that the requirements of the law and associated regulations are met. This means regular monitoring of all plants that fall under these regulations. Ideally, monitoring is carried out by spot measurements of source emissions at unannounced times by officers of the agency. In practice, spot testing with adequate frequency may be too expensive, and a compromise involving some agency testing combined with self-monitoring requirements on individual polluters will be used. In the most sophisticated systems, continuous stack monitoring equipment is specified, possibly linked directly to the control agency. Continuous monitoring of this type is only an economic proposition for major sources who individually make a significant contribution.

For the individual sources whose contribution is significant only when they are considered collectively, a different approach to policing is required. The most important source in this second category is the private motor car – the most significant contributor of the photochemical smog precursors NO_x and hydrocarbons, and of carbon monoxide in urban airsheds. The approach most frequently taken to automotive emissions control has been to require the manufacturer to meet set standards on all new cars and to test vehicles on the road randomly to ensure they are meeting the standards and have not been tampered with. It has not been common for retrofit of control devices to be required on cars whose date of manufacture predates the regulations. In practice, the methods have not proved particularly successful, new vehicles often failing to meet the requirements when independently tested and the probability of on-road testing proving too low to deter tampering.

The second function of the control agency, to determine the effectiveness of the measures taken to abate pollution, involves ambient air monitoring of selected pollutants. The data obtained can then be compared with the predicted values based on pre-control concentrations and known emissions and meteorological information, and action recommended accordingly. The same monitoring data, alongside effects observations, are also necessary for establishing air pollution criteria. No matter what use the data are put to, it is always necessary to ensure that the sites chosen for taking ambient measurements are not directly affected by any single source or unduly influenced by any physical features such as nearby buildings. In practice, the number of sites available that both fulfil the above requirements and have the necessary security and power supplies is strictly limited and some compromise is necessary. Within a single jurisdiction, and preferably within all areas where any cross transport of pollution occurs, instrumentation and procedures should be standardized and strict guidelines followed for site selection to ensure the values obtained are as close to a true representation of atmospheric concentrations as possible. The number of permanent and temporary monitoring stations in any area is a function of balancing costs with the seriousness of the problem, the complexity of the terrain, and the size and density of the population, or the value of the crops or amenity to be protected. Ultimately, this is a political decision reflected in the level of fund allocation to the management agency.

In the longer term, it may not prove possible to maintain air quality and accommodate growth by the application of emission controls only. The air pollution implications of future planning options should, therefore, always be an input into the decision making process. Even in circumstances where there is no risk of exceeding air quality goals, a proper weighting of air pollution implications could influence the choice towards options that maintain or possibly improve air quality without detracting from the original scheme's intentions. The necessary liaisons between the various government bodies responsible for growth and development and the government agencies whose job has been to alleviate the unwanted side effects of such growth, has been singularly lacking in the past. Recent legislation in the USA and

Australia has gone some way towards redressing this by requiring environmental impact statements to be provided for all projects involving government finance, thereby helping to ensure that the air pollution outcome is at least a consideration.

Evaluation of planning options has as its basis the comparison of the total emissions associated with each option, and the associated concentration distribution of pollutants throughout the airshed. Existing emissions are obtained by making an inventory of all sources integrated over some chosen period of time to remove short term fluctuations. To be accurate, the inventory should be based on the measured emissions from all sources; this task increases in formidability as the number of sources increases, and some approximations are required. A large amount of reasonably accurate emissions data will be available for controlled sources and this can be used directly. Data for uncontrolled sources has to be estimated from the number of sources in each uncontrolled category, together with emissions data from typical sources of each type, and appropriate scale factors based on fuel usage or material throughput of some sort. Automobile emissions are a special case and are generally based on average kilometres travelled, the age of the vehicle, the measured average emissions for a vehicle of that age on some standard driving cycle, and the number of vehicles in each age category. Projected emissions for future proposals contain more uncertainties and have to be based on standard emissions data for each source category, combined with any proposed future controls and estimates of such things as future levels of vehicle ownership and usage, which will reflect unknowns such as fuel prices. Emissions data on their own give some idea of the comparative level of pollution from individual proposals. An even better comparison can be obtained by using a model to calculate atmospheric concentrations for each of the options and the distribution of these concentrations spatially, that is if a suitable validated model is available.

There are several ways in which the planning process can be used to protect air quality, including the use of green belts and sites of high dispersive characteristics as discussed earlier. For airsheds that are at or near their maximum capacity for accepting pollutants, measures have to be taken to limit new industry if the possibility of tighter emissions control is not available. Two approaches are possible. First, to prohibit future development in certain areas, and second, to offer economic incentive to industry to locate elsewhere. The latter could take the form of direct financial aid or of tax relief, or a financial disincentive such as making the level, and therefore the costs, of control so high for new industry that it is forced to locate elsewhere.

New housing and urban renewal schemes can and should be planned to minimize air pollution. For example, high density centralized housing has less of a pollution potential than the suburban housing sprawl because it offers the possibility of more efficient and therefore less polluting heating systems, and also of more viable public transport networks. The transportation area is another where careful planning can have a significant impact on air quality, such as measures to encourage use of public transport or small cars, or even multiple use of cars. Transportation options are discussed in more detail in Chapter 7, which deals with the air pollution implications of future energy use patterns.

Finally, part of any air pollution agency's brief should be to ensure due consideration is taken of environmental matters in future planning. An air pollution agency is in a good position to speak authoritatively on planning issues and to apply pressure to obtain the most desirable outcome. The effectiveness of such pressure is greatly aided if the public is provided with clear information regarding the issues involved as part of a continuing programme of public education. An informed public, aware of the trade-offs involved in the decision making process, is ultimately the best insurance that environmental matters are given their due weight.

7 The Future

Air pollution is primarily a result of man's consumption of fossil fuels, a consumption that is of relatively recent origin. The present high level of fossil fuel usage cannot be extended indefinitely into the future because the source – laid down over many thousands of years – is finite. Fig. 7.1 shows just how short this period of fossil fuel usage will be in the history of mankind. Future energy use patterns are, as a consequence, likely to be very different from present patterns, both in terms of the use of new sources of energy and the deployment of existing sources. These changes will alter the nature of air pollution problems. We therefore need to consider what our future options for meeting our energy requirements are, and what implications these options will have for air quality and air pollution control. Alongside this, consideration must also be given to changing community attitudes towards air quality which may require increasingly stricter controls on new and existing sources.

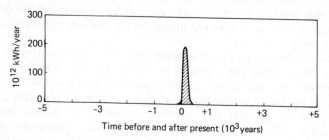

Fig. 7.1 World consumption of fossil fuels on a time scale of 5000 years before and after the present.

Domestic and Commercial Heating

In those parts of the world occupied by many of the more highly developed nations, there are considerable periods each year during winter when it is necessary to heat the places where people live and work. Much of the traditional air pollution in Europe and North America – the smogs of London, Pittsburgh and elsewhere – was primarily due to heating methods, particularly to the incomplete burning of poor coals in badly designed grates and stoves.

(Reduction of domestic air pollution, which has been most successful in the two cities mentioned, has been carried out by replacing the traditional fuel (coal) with other types, mainly natural gas and electricity.) In England, where until recently natural gas was not available, the combustion of specially prepared solid fuels (coke, Coalite and others) in better designed grates proved a low cost answer. Even so, replacement of the old 'open fire' with better systems was only made possible by large subsidies from government authorities. In the United States and Canada, the widespread use of natural gas has helped to maintain low levels of air pollution from home heating.

Natural gas is a particularly clean fuel, convenient to use, and easy to transport, but supplies are more limited than those of the solid fuels it has replaced. The possibility exists of extending gas supplies by synthesizing it from solid fuel, namely, coal, as was frequently done in the past, and pipelining the product from the coal fields to the urban centres. In recent years, various processes have been developed for the complete gasification of coal, without by-product coke. Several of these, notably the Lurgi process developed in Germany, have been in operation in several countries (Germany, South Africa, Australia). They can only be used with a limited range of coals, and new processes are under development which will handle more coal types. The gas, after purification, can replace natural gas as a source of direct heat, although it has a lower heating value. It may also prove possible to synthesize the gas within the underground coal seams and avoid the mining step altogether.

Electric heating produces no waste emissions at the point of use and, as electricity generating stations are mostly located away from cities, no urban pollution. However, as we have already seen, power stations, particularly coal burning power stations, are major polluters locally and possibly globally through their emissions of carbon dioxide, sulphur dioxide, nitrogen oxides, and particulates. Conversion of coal to electricity is also very inefficient, only about 35% of the energy in the fuel being turned into electrical energy with an additional 5% being lost during transmission to and transformation at the point of use. These disadvantages are offset by the use of low grade coals which are not in high demand for other purposes, and have unacceptably high emissions when burned directly for domestic heating. A further advantage can be gained by using electricity only at off peak periods for heating applications, thus allowing for a more even base load at the power station.

Light fuel oil has had extensive use in Europe and North America as a heating fuel, although in recent years there has been a trend towards other forms of heating, mainly as a result of increasing oil costs. It should, however, be noted that it is much more efficient to burn oil directly for heating than to burn it in a power station for electricity generation.

In densely populated areas, with large apartment house complexes – New York, Stockholm, London, Amsterdam and some German cities – greater efficiency is achieved by district heating from a central boiler house. It has sometimes proved possible to use the heat from large, centrally situated garbage incinerators for this purpose, as is done in parts of Stockholm.

An important adjunct to using only the most suitable energy for heating purposes is conservation. The conservation approach equally saves energy and reduces pollution. For example, heat losses from buildings are direct energy losses and these can be greatly reduced (by at least 10%) by insulating walls, floors and ceilings and by double glazing windows. Other elements of house design can also conserve energy, for instance catching the winter sun and excluding the summer sun by utilizing deep eaves over south facing windows in the northern hemisphere, and north facing windows in the southern hemisphere. The eaves could be supplemented or replaced by deciduous trees, which provide shade in summer but not in winter. Further savings can be made by persuading people to tolerate slightly colder conditions indoors in winter – such a measure greatly reduces the heat flow to the outside by lowering the indoor/outdoor temperature differential. A change in average indoor temperature of 1°C can be readily accommodated by wearing warmer clothing and provides a significant saving over a large number of households. The same measures are equally relevant to areas that use air conditioning in the summer.

So far, we have only considered ways of extending or conserving fossil fuel supplies. Another possibility particularly appropriate to home heating and hot water supply is the use of renewable forms of energy. Solar energy is a strong contender in this area, coupled with a heat storage system to take account of the fluctuations in the amount of sunshine received. Heat can be collected using relatively simple commercially available devices, generally a black radiation absorber that directly transfers the heat to a gas or a liquid. The heat is then stored for liquid systems in an insulated tank of the liquid, and for gases in high thermal capacity materials such as rocks. It is also possible to use certain salts for storage. These undergo a phase change at a suitable temperature, for example, hydrated sodium sulphate $Na_2SO_410 H_2O$ which dehydrates at 32.3°C absorbing heat and readily releases this heat on rehydration. Systems in present use provide a proportion of the heating requirements and use auxiliary energy for the remainder, thereby balancing the capital cost of solar collectors and storage with the benefit gained.

Even our fossil fuels are solar energy-stored photosynthetic products of past ages laid down without oxidation until present generations decided to burn them. The biosphere similarly is the stored products of photosynthesis, and as such has a potential use as fuel. The biosphere has the further advantage of being renewable. Wood, for instance, can be used directly for home and water heating. It can be argued that wood can never be more than a minor contribution to the world's needs, applicable only to small towns and rural communities. Nevertheless, it is a contribution that is important now in many parts of the developing world with the possibility of remaining so in the future. Some industrial uses of plant material for fuel also exist in natural product industries where it is a waste product, such as in the timber and sugar industries. Wood burning contributes its own quota of pollutants to the atmosphere including particulates and polyaromatic hydrocarbons, carbon dioxide, carbon monoxide and sulphur dioxide, so that the replacement of coal or oil by wood is not an advantage in pollution terms, and the replacement of gas by wood is a regressive step.

In areas of geothermal activities – New Zealand and Iceland – steam from geothermal bores has been used for heating. The distribution of active geothermal areas throughout the world limits the applicability of the method; however, there is a potential for the related possibility of pumping water through hot sub-surface rocks, where these occur reasonably close to the surface. The use of geothermal heat cannot be considered non-polluting as the steam is often accompanied by gaseous emissions, including hydrogen sulphide and radon, both of which must be considered hazardous.

Electric Power Generation

Our modern society depends on electricity as its primary source of readily available energy, and most of our electricity is generated from non-renewable resources. Electricity is not only used in the home for light and heating, but in industry for processing. The recovery of aluminium from the oxide (alumina), and the electrolytic purification of copper, lead and zinc use very large quantities of electricity. In many developed countries, the demand for electricity has been doubling each decade and although some slow down in this rate is now apparent, we have become dependent on electric household aids, and the industrial power demand has been rising.

Electric power generation from coal and oil is the world's major source of pollutant sulphur dioxide and carbon dioxide, and is also a significant source of nitrogen oxides and particulates.

We have already considered the rationale behind the use of coal fired generating stations; specifically, that the coal is a low value fuel in reasonably abundant supply that cannot be cleanly used as a direct source of energy in the urban situation, and it is therefore best to burn it, albeit inefficiently, to produce a secondary product, electricity, that can be transported and cleanly used in out cities. The use of gas cleaning equipment on power stations flues is intended to reduce the concentrations of pollutants to a level that can be safely dispersed to the atmosphere without producing any direct damage. This view does not take into account the long term transport of sulphur dioxide and the observed increasing acidity in rainfall, nor does it consider the possible effects of the global increase in carbon dioxide concentration, both a function of power station emissions. The answer to the sulphur dioxide problem would seem to be a reduction in electricity usage, improved controls, or the development of alternative means of electricity production such as nuclear power. The first two alternatives will be insufficient in the view of the US EPA, who have estimated that even with the best available means of control and conservation there will be an increase of around 10% in the US sulphur dioxide emission between 1975 and 1995. The nuclear alternative, as we shall see, has particular problems of its own but even if there was no barrier to a nuclear programme, lead-in times are such that it is unlikely to have any significant effect on pollution levels before the turn of the century.

The situation with carbon dioxide parallels that of SO_2 except that CO_2 is not amenable to control measures. It is, therefore, unlikely that any change in the present observed pattern of atmospheric carbon dioxide increase will occur in the near future. The most likely outcome will be an adjustment on the part of populations to any climatic changes that might result from the expected increase.

Alternatives must, of course, be found for the long term fulfilment of energy needs. Renewable resource alternatives such as solar, wind and wave power are attractive from a resources and pollution point of view but, as yet, cannot compete economically except in remote areas. Solar energy can be used to make electricity and in theory and amount available is large, 1.36 kW m^{-2} being the quantity external to the earth's atmosphere. It has been estimated that, without the atmosphere, if this energy could be fully utilized it would supply the annual energy consumption of the United States in less than a day. In fact, only about half this incoming energy reaches the earth's surface, the amount per unit area varying with geographical location and season of the year. The amount of this energy that can be converted into electricity is a function of the efficiency of the collection and conversion procedures which at present is not very high.

In practice, solar radiation can be converted into electrical energy either indirectly or directly. Indirect conversion can be carried out by concentrating the radiation using sun-following mirrors to convert water to steam, and using the steam to generate electricity by conventional means. Such a system will only operate in direct sunlight. This means the generation will be intermittent and the collecting area required to produce a given amount of electricity will vary with the intensity and duration of sunlight for the area concerned. It is estimated that for a hot dry region such as Western USA, North Africa or Central Australia, such a power plant with the expected efficiencies of conversion would require 13 to 25 km^2 of collectors to provide 1000 megawatts of power. This is larger than the area of a conventional power plant but less than that required for a coal fired plant and its associated open cut mine.

Direct conversion to electricity can be achieved using the photovoltaic effects. Cells made up of specially prepared semiconductor materials, such as silicon, develop a voltage difference across their crystal surfaces when exposed to direct sunlight and hence are a source of electrical energy. Such systems have the advantage of being equally efficient used in small arrays to provide power for a camera or a house, or in bigger arrays for large buildings. They are, however, expensive, inefficient (around 15%), and require an associated storage system, generally batteries, to provide a continuous power supply at night and on days of low sunshine.

Battery storage systems, which are both expensive and of limited life, are a feature of all intermittent solar derived electric power systems, including wind power. A method that has been suggested to use solar derived energy without the need for storage systems is to convert the temperature difference between the surface and deep oceans into electrical energy. A trial plant based on the ocean's temperature gradient is expected to be commissioned in the USA in this decade.

Energy is also available from tidal and wave power. Tidal power can be harnessed by using the rising tide to fill a reservoir which is closed by a dam. Outflow from the dam at low tide can then be used to power a turbine in much the same way as is used for hydroelectricity. There are only a limited number of sites worldwide where the difference in height between high and low tide is sufficient to make a tidal power station a reasonable proposition. One such situation is on the Rance estuary in France, where a tidal station is at present in operation. There are also proposals for capturing the considerable energy in waves. This is, however, a more difficult problem because the energy is spread over so wide an area.

Overall, it would appear that the use of renewable resources for electric power generation is a useful supplement and extender of our present methods of generation only in special situations where economic conditions warrant it.

Geothermal fields are being used to generate electricity in Italy (Larderello), the United States (Sonuma County, Northern California), New Zealand (Wairakei), Iceland, Mexico and the Soviet Union. In these fields a bore, about 1000 m deep is put down and a mixture of water and steam under pressure comes up, which is then led to turbines in the generating station. The water and steam are separated in cyclones, and the high pressure water can be 'flashed' into steam and used for more electricity generation. However, the steam pressures are comparatively low when compared with large modern thermal power stations, and this results in large turbines with limited generating capacity. Still, it must be remembered that the 'fuel' is quite free, and the resultant cost of power is low. But little is known about the 'life' of a geothermal field, and so while geothermal energy is produced at low cost, long term prospects are unknown. It can only supply a small fraction of the energy demand even in those countries where it is available, and as we have already seen, it is not without pollution problems.

There are also other methods of producing electrical energy from fossil fuels which do not use the conventional turbine to generate electricity. Two which have attracted much interest in recent times are 'magneto-hydrodynamics' (MHD) which converts heat directly to electricity, and 'fuel cells' which rely on the direct conversion of chemical energy. In MHD an electrically conducting gas at very high temperatures is forced through a duct at high speed in the presence of a transverse magnetic field. An electromotive force is induced in the gas, and current is extracted by electrodes and delivered to an external load circuit. The gases leaving the MHD system are still very hot, and can then be used for conventional forms of electricity generation. The overall efficiencies in turning the energy in a fossil fuel to electricity in such a system would be 50% higher than for conventional power generation, but there are so many technical difficulties associated with the very high temperatures that MHD does not seem a feasible means of power generation in the foreseeable future.

The fuel cell is essentially a continuous battery into which reagents are fed; the electricity is produced chemically and not by physical means. In a sophisticated system intended for base load power generation, gases from coal gasification could be turned directly into electricity with an overall efficiency of over 60%, almost twice that of conventional power generation. Except in special circumstances, the potential of the fuel cell has not been realized because of the small currents that can be obtained per unit electrode area, and because of the expense of the electrodes required to obtain higher current densities.

Among alternative power sources are two which are fully developed technically: hydroelectric power, and some forms of nuclear power. In hydroelectric power generation, electricity is obtained from the energy of water flowing from a higher level to a lower one, and turning a turbine in the process. Some countries, such as New Zealand, Switzerland, Norway and Canada, obtain virtually all their electrical energy from this source, to such an extent that electrochemical industries are based on hydro-power. Hydro-power also has the advantage of no air pollution and it can be readily controlled by turning on a motorized valve in the water supply, so there are no problems associated with peak loads. There are, however, a number of difficulties in the widespread development of hydroelectric resources. Invariably, it requires the damming of large volumes of water, flooding valleys and covering vast areas of land, often valuable for commercial or recreational purposes, or undisturbed 'wilderness' areas in which there will be undesirable environmental changes. This has raised widespread protests in some instances, but with limited success. Notable have been the raising of the levels of Lake Pedder in Tasmania, of Lake Manapouri in New Zealand, and the flooding of the 'Rainbow Bridge', a natural sandstone formation in Southern Utah.

Because of the waste in using fossil fuels for dealing with peak power demand, the development of hydroelectric resources has been favoured by electricity generating organizations. A recent idea is 'pumped' storage, a method that could also be used with solar or wind generated electricity, or with direct windmill pumping. Water, after passing through the turbines, is held in a large storage pond. During the night, off peak power from base load thermal (or nuclear) stations is used to pump the water back to the upper level. This uses the otherwise idle or unused capacity of the base load station which is often nuclear (see below). At present, the 'peak demand' for electricity is frequently supplied by bringing into action older power stations, sited in cities, which tend to have higher levels of pollutant emissions, and use fuels such as gas and oil which would be better used for other purposes. The replacement of these by pumped storage has much to recommend it, both on conservation and air pollution grounds.

Nuclear reactors at present produce around 13% of the total electricity production in the USA and 10% of the production in the UK. The capacity exists for this percentage to increase; however, because of uncertainties associated with nuclear power production and waste disposal of radioactive materials the level of public acceptance has been low and growth has not been what was originally predicted.

Nuclear energy derives from the conversion of mass into energy, in accordance with Einstein's law $E = mc^2$ where E is the energy, m the mass converted, and c the velocity of light in a vacuum. Most existing nuclear power plants derive their energy from the fission of the uranium isotope, uranium 235 (^{235}U). The heat released by the fission is used to convert water into steam and operate a turbine to generate electricity in much the same way as in a fossil fuel power station. Several different reactor technologies are in current use, all with efficiencies of the same order (around 35%) as conventional thermal power stations. Their viability in particular situations depends on: the cost and availability of other fuels, the level of substitution of other energy sources for heating, and increasingly on the acceptability of nuclear stations to the general population.

Two types of reactor in current use, the boiling water reactor (BWR) and the pressurized water reactor (PWR), use water as moderator and coolant. For this reason they are classified under the general heading of light water reactors (LWR). Uranium 235 fission requires the capture of thermal neutrons and the presence of the water lowers the energy of the neutrons released in the fission process, so that they in turn can be captured by a ^{235}U atom which leads to a self-sustaining fission chain. The hydrogen in the water also captures some of the neutrons released, and to compensate for this and maintain criticality, the uranium ore must be enriched in ^{235}U from its normal value of 0.72% to between 3 and 5%.

Enrichment is carried out by gas diffusion. First the ore is purified and converted to uranium oxide U_3O_8 and then to the corrosive hexafluoride UF_6. UF_6 is passed through a multi-stage gas diffusion system in which the predominantly $^{238}UF_6$ gas is enriched in the lighter and more readily diffusable $^{235}UF_6$. The enriched UF_6 is then converted to UO_2 which is used to prepare the fuel rods. In both light water reactor types the fuel rods consist of uranium oxide (UO_2) pellets sealed in a zirconium alloy tube. Zirconium alloy is chosen because of the low capture probability of zirconium for thermal neutrons and the high resistance to corrosion of its alloys.

Fig. 7.2 is a simplified representation of the boiling water and pressurized water reactors (BWR and PWR). In the BWR, water is pumped through the reactor core, where it boils and produces steam which operates the turbines. The steam thus generated has a pressure of 6.8×10^6 N m^{-2} and a temperature around 285°C. In the PWR, the water passing through the reactor core is initially pressurized to 13.6×10^6 N m^{-2} to prevent boiling. This water, after leaving the core, passes through a steam generator where its heat is transferred to a secondary water system which produces the steam at around 260°C which drives the generator. A third type of reactor uses a gas such as helium to cool the reactor and transfer the heat to the generating system. The hot gas, possibly as high as 760°C, transfers its heat to a secondary water system without direct contact in an analogous way to the PWR reactor.

In Canada heavy water reactors are in operation (CANDU reactors). These use deuterated water ^2H_2O as the neutron moderator. The advantage here is that deuterium has a much lower capture probability for thermal neutrons than hydrogen, and natural (unenriched) uranium can be used as the fuel. The cost of extracting ^2H_2O from water is traded against the cost of enriching the uranium. The heavy water can also serve as a coolant in which case heat is transferred to a secondary water system as with the PWR.

During the fission chain reaction, a small fraction of the major uranium component of the fuel^{238}U is converted to plutonium 239

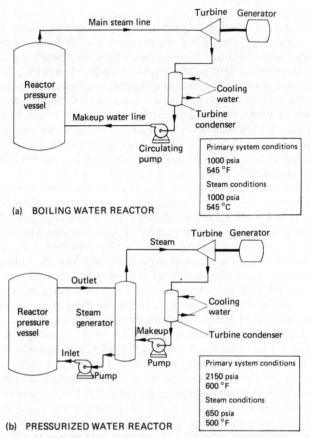

Fig. 7.2 Schematic of boiling water reactor and pressurized water reactor. (From Strauss, W. (ed.) *Air Pollution Control – Part II*, Wiley Interscience (1972).)

(^{239}Pu), a fissionable product that can contribute to the power output of the reactor. This conversion of a non-fissionable material ^{238}U to a fissionable material is the basis of the proposed fast breeder reactors. This would seem to be a necessary step if nuclear power stations are to play a role in providing future electricity supplies, as it has been estimated at current rates of growth ^{235}U supplies are only sufficient for another 20 or so years. Use of ^{239}Pu breeder reactors could extend the use of fission reactors for several hundred years. However, the exceptionally long half life of ^{239}Pu of 24 390 years and its potential use in nuclear weapons have caused serious objections to be raised to the building of plutonium breeder reactors, although it is claimed that plutonium is sufficiently denatured by leaving it in contact with its own fission products. Alternative breeder reactors using uranium 233, which can be obtained from thorium 232, also extend the potential lifetime of fission reactors without the same safety problems as the recovered uranium 233 from this process can be suitably denatured with ^{238}U and used as a fuel in the light water reactors already discussed.

In the breeder reactor fast neutrons from the fission of ^{235}U are used to convert ^{238}U to ^{239}Pu (Fig. 7.3). In operation, more ^{239}Pu is produced from ^{238}U than is used up in the energy producing fission of ^{239}Pu. The use of non-thermalized neutrons favours the conversion of ^{238}U over the fission of ^{235}U and minimizes the non-fission absorption of neutrons by structural materials. Further reduction of non-productive absorption is necessary and this is done by using a non-moderating reactor coolant. The coolant used in existing prototypes is liquid sodium which has a low capacity for absorbing and thermalizing neutrons and ideal heat transfer properties. All the same, some neutrons are captured by the coolant, resulting in the formation of the radioactive isotopes of sodium ^{24}Na and ^{22}Na. In addition, fission products leak into the liquid sodium or are vented there intentionally. The sodium is as a result highly radioactive and a second liquid sodium circuit isolated from the first is required to transfer the heat to a water circuit producing steam to drive the generators. Liquid metal fast breeder reactors can operate at higher temperatures and therefore at higher thermal efficiencies than light and heavy water reactors. Their overall efficiency is expected to be around 40%, which is just above the upper limit of the best conventional fossil fuel fired stations.

The nuclear power industry consists of not only the power station itself but the mining, milling, conversion and enrichment processes

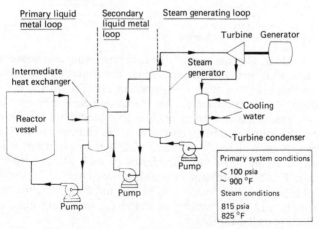

Fig. 7.3 Schematic of the liquid metal cooled fast breeder reactor. (From Strauss, W. (ed.) *Air Pollution Control – Part II*, Wiley Interscience (1972).)

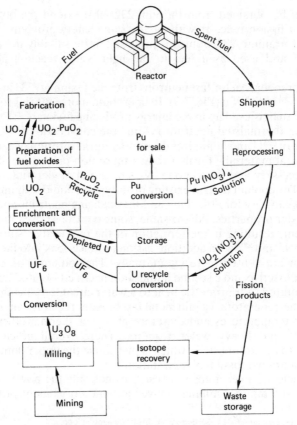

Fig. 7.4 The steps in the uraniun cycle. (From Eicholz, G. G., *Environmental Aspects of Nuclear Power*, Ann Arbor Science (1976).)

that precede it, and the transporting, reprocessing and waste storage operations that deal with the spent fuel and contaminated plant. The steps in this uranium cycle are shown in Fig. 7.4. Not every step is potentially an air pollution problem although each may contaminate other sectors of the environment. It is estimated that emissions to the air from the mining, milling, conversion, enrichment and fuel fabrication processes will result in doses of less than 10 mrem a year to individuals living near the boundaries of such plants – this is not considered to be significant in health terms. Doses to employees in these plants would be considerably higher and the association between lung cancer and the breathing of radon gas in underground mining operations is well established.

It is the step where new radioactive products are produced that introduces the main hazards associated with the industry, both in the power plant itself and in the treatment and disposal of the unwanted wastes from it. Radioactive materials are produced in the reactor by the fissioning of the fuel. The products of the fission of uranium include radioactive isotopes of krypton, xenon, caesium, rubidium, barium, strontium, iodine and bromine, and the isotope of hydrogen, tritium. Further products are generated in the cooling water including isotopes of argon, fluorine, nitrogen and oxygen, and another range of radioactive isotopes is created within the structural materials surrounding the core. Plant gases are emitted to the atmosphere generally after a holding period of up to 30 days, which allows the short lived isotopes to decay. The main source of radioactivity in the emitted gases is krypton 85 (^{85}Kr), together with smaller amounts of tritium (^{3}H), and iodine 131 (^{131}I). In normal operation, the level of radioactivity in the released gases is strictly controlled. Current US regulations require that the average dose at the plant's boundaries be no more than 5 mrem per year, which is expected to produce negligible health effects. Other problems are associated with the cleaning of the cooling water, which may contain contaminants leaked from the core, including tritium, which can diffuse through metal barriers at high temperatures, and also the disposal of contaminated old plant. Neither of these are air pollution problems but their release to the environment can add to the average dose that an individual might be receiving, and in some cases radioactive materials released to one part of the environment may well find their way into another. It is suggested that this may be the case at Sellafield in the UK, where material released to the estuary is transferred to the atmosphere as an aerosol formed by wind action on muds exposed at low tide. Again, under normal conditions, concentrations are such that adverse effects are not expected, the 5 mrem dosage limit per year being the dose from all sources.

The biggest cause of concern with nuclear plants is the possibility of unintended release of radioactivity as a result of plant malfunction or of accident. The level of safety precautions built into a nuclear power plant is high and the probability of serious accident claimed to be low. However, there is not universal confidence in the probability calculations – a lack of confidence that has been reinforced for many by the recent failure of the safety mechanisms at Three Mile Island in Pennsylvania. The most serious type of accident that could occur would involve release of radioactivity from the reactor core as a result of a failure in the cooling system and the resultant meltdown. Various calculations have been made of the probability of population damage based on the probability of a core meltdown and the probable level of emissions from a meltdown or other power plant accident. Such a calculation is the US Nuclear Regulatory Commission Report of 1975, which estimates an early fatality risk for a hundred nuclear reactors in operation in the United States. This is not only less than the risk from automobile and aeroplane accidents but also less than uncontrollable risks such as hurricanes and earthquakes. On the other hand, risks of nuclear accidents assume greater significance for many people than the risk factors would justify, for several reasons. First, the calculation of risk is not based on existing statistical data and therefore is more suspect than the risks used for comparison, which are. Second, things like automobile accidents are often seen to be within the control of the individual and natural disasters are seen as Acts of God; nuclear accidents fit neither of these categories and can be seen as risk imposed by society.

Other risk areas associated with nuclear power plants concern fuel reprocessing, waste disposal and the possibility of nuclear materials being stolen and used for nuclear weapon fabrication.

Radioactive waste from nuclear power stations must be stored, and for the long-lived radionuclides this storage must be secure for many thousands of years. Various methods have been suggested; solidifying the waste in a glass and burying it in a stable geological structure with good heat transfer properties appears to be favoured at present. Reprocessing is not as yet a major activity associated with nuclear power plants but will become increasingly important in the future. Basically, it is the step where uranium and plutonium are separated from the spent fuel rods for re-use, and the waste materials – the rod cladding and fission products – are prepared for storage. It is, therefore, a step where the probability of release of radionuclides to soil and water is high under normal operating conditions because it cannot be operated as a closed system, as is the case with the power plant itself.

In summary, under normal operation nuclear power plants produce no significant air pollution. They are capable of supplying many future energy needs, replacing high pollution fuels and also freeing these for use as feedstocks for other industries, e.g., plastics, drugs and fine chemicals, or for use as or conversion to transport fuels. On the other hand, the probability of accidental release is still in doubt and the level of public acceptance of such plants is sufficiently low to have significantly delayed the nuclear power programme in both Europe and North America. In addition, uranium supplies, like coal and oil, are limited making this too an option which will ultimately be exhausted, probably within a few hundred years.

Nuclear fusion is potentially a source of unlimited energy in the future. The fusion reactions of interest are those involving the heavy isotope of hydrogen, deuterium (^{2}H), which occurs naturally in sea water in the ratio of 1 deuterium to every 6500 hydrogen atoms. The possible reactions and their resultant energy release are as follows:

$$^2H + {}^2H \rightarrow {}^3He + n + 3.27\,\text{MeV}$$
$$^2H + {}^2H \rightarrow {}^3H + {}^1H + 4\,\text{MeV}$$

$$^2H + {}^3H \rightarrow {}^4He + n + 17.6\,\text{MeV}$$

where n is a neutron, ^3H the even heavier isotope of hydrogen, tritium, which would otherwise have to be made by neutron bombardment of lithium, ^3He and ^4He are isotopes of helium. It is estimated that if these processes could be successfully controlled, there is as much energy in 1 cubic metre of sea water as in 270 tonnes of coal or 1360 barrels of oil.

The required temperature for the fusion reactions is of the order of 10^8 °C and this is where the problem lies, both in finding the means of attaining such temperatures and also in confining the reaction under such extreme conditions. In theory, the air pollution hazards of fusion reactors would be associated with the release of tritium gas only, and consequently would be of much less consequence than those of the fission reactors. However, problems would still arise from induced radiation in the plant. This could possibly be released if an accident occurred and would pose disposal problems when the plant became obsolete.

Fusion offers an energy supply of great potential but its future is at present uncertain and even if its development becomes possible the lead in time will take several decades.

Transportation

In the overall pattern of pollution, the private motor car produces more than one third of the major gaseous air pollutants in our cities. This is added to by our other means of transportation, trucks, buses, trains and aeroplanes. Control devices, combustion modifications, and alternative engine designs all go some way towards reducing this pollution burden (Chapter 5), in most cases adding an energy penalty (with the notable exception of the diesel engine). Mobility has become an intrinsic part of 20th century life – a mobility that is almost completely dependent on supplies of liquid petroleum fuels, the least abundant of all our fossil fuel sources. We obviously cannot continue to rely on this source of energy in the future to supply our transport needs. Future options available include obtaining liquid fuels from other sources such as coal conversion or oil shales, using alternative means of powering vehicles, using less power to achieve the same travel requirements or, as is most likely, some combination of all three.

Each option has different air pollution implications. Obtaining liquid fuels from other sources, for example, leaves us with our automobile pollution problem as before but adds to it a further problem of the pollution produced by the conversion or extraction processes. Like coal fired power stations, liquid fuel production plants would be at the site of the raw materials – coal or shale oil – and therefore away from major centres of population. The main contribution of such plants, if they are properly controlled, would therefore be to regional air pollution problems such as acid rain and increasing air opacity due to particulate concentrations. Alternative fuels such as liquid petroleum gas (LPG), liquified natural gas, ethanol, methanol and methane, are more volatile than conventional fuels and have lower emission factors. LPG and natural gas, however, limit the range of vehicles to between 50 and 160 km, making them most suitable for city driving. In addition, both are fossil fuels and limited in supply. Methane, methanol and ethanol, on the other hand, can be made from renewable resources – from wood, waste organic material and the fermentation of crops, making them suitable contenders for supplementing existing fuel supplies. Although renewable fuel sources are currently under investigation in several countries, to date only Brazil is operating a commercial system using a 20% ethanol/petrol mixture, which requires no engine conversion on standard vehicles. The goal of the Brazilian government is to be 40% self-sufficient in fuel by 1985. This will mean more conversion to pure alcohol vehicles – a programme which is in its early stages at present – and further cultivation of the crops from which the alcohol is produced, sugar and more recently cassava. These crops are particularly suited to Brazil's climate. It is unlikely that appropriate raw materials can be produced in sufficient quantities world-wide, but for those countries that can grow suitable crops in excess of their needs alcohol is a valuable supplementary fuel.

Two other alternative power sources for transportation are also receiving considerable attention, namely batteries and fuel cells. The problem with both these is that, although the vehicle is virtually pollution free, either the available power and therefore speed and acceleration, or the available range, is limited. Batteries are capable of supplying sufficient power but the range of practicable batteries is limited, making them useful only for the short city trips they are used for at present. Fuel cells, on the other hand, have longer range but insufficient power to move the vehicle at a reasonable speed. The manufacture of both fuel cells and batteries requires both energy and high cost raw materials, so that although improved vehicles using these power sources alone or in combination are feasible, their cost at present rules them out. Any massive conversion to battery powered vehicles requires an additional amount of electric power for recharging, and this requires increased power plant production, with its associated pollutant emissions. It can be argued in this case that it is easier to control emissions from a large single source like a power station than from a multitude of smaller ones.

The final option is basically that of conservation, namely, to use less fuel for transportation purposes. Moves in this direction, like the search for alternative fuels, are chiefly motivated by the desire to conserve liquid fuels and overseas exchange. One beneficial spin-off, however, would be a reduction in air pollution.

A considerable amount of fuel conservation is achievable without any major change in present life styles by maintaining the private automobile but making it more efficient, in particular by moving to smaller, lighter cars. Further savings could be achieved by successfully encouraging people to use more public transport, particularly for the regular journey to and from work. Such transport, to be acceptable, must be fast, frequent and reliable. Public transport systems using electric power are available that are virtually pollution free. Many urban train systems fit this category, but electricity is also applicable for trams and trolley buses. As has already been noted, any major move to electricity will result in additional power station pollution; easier to control than a large number of cars but posing slightly different problems. Unfortunately, the design of many modern cities, particularly those developed in the last 30 years, has made the car a necessity and public transport a very expensive alternative. Supplies of food and other essentials are only available in locations perhaps 2 or more kilometres from residential areas. Large shopping complexes, frequently protected from the weather, have grown up on the outskirts of our cities, with vast car parks. New schools, places of entertainment and even universities, have followed the same pattern.

Public transport is most viable in densely populated areas, as is found in older European cities where the usage is high and the distance of travel relatively short. It is favoured more if the usage can be evenly spread throughout the day, avoiding rush hour situations. This is partly achievable with staggered or flexible working hours. Structural changes in existing cities are unlikely to occur without a considerable amount of economic pressure, such as might result from escalating petrol prices. One partial solution for widespread cities is the encouragement of car pooling by providing, for example, special driving lanes that can only be used by cars containing three or more people. Another approach is to encourage use of a disperse public transport system by providing parking lots at suburban transport stations, to encourage people to limit their driving to and from the nearest of these. This latter alternative is aided by a shortage of parking in the central city. Such measures, where they have been tried, have proved only partially successful.

Changes have occurred with intercity transport where the private car and the aeroplane have replaced the train in countries like the USA, Canada and Australia, where distances are large and population densities small. In countries like Japan, or in Europe where intercity distances are smaller, trains still remain a convenient means of travel and their continued use has led to the development of super fast trains providing a faster service than could be achieved by car and, in many cases, than by aeroplane, particularly when account is taken of the journey to and from the airport. The effect on air pollution of different methods of moving passengers intercity is likely to be small; the sources are dispersed and away from urban and industrial centres. Movement of freight, though expensive in energy terms if moved by road, as it increasingly is, would similarly not be expected to have major air pollution implications.

Industrial Pollution

Industrial sources of air pollution were the first to be recognized, segregated and to some extent controlled. Many of the control methods are described in Chapter 5, but control of pollution from these sources still presents great technological challenge if any industrial growth is to occur without deterioration in air quality. New developments in control methods, alternative processes that are more easily controlled and less polluting, alternative sources of energy where applicable, including desulphurized fuels, can all contribute to cleaner air without stagnation of industry and commerce. Increasing populations, economic growth in the underdeveloped world, and the demand for a higher quality and expectation of life, are all pressures towards alternatives that are less polluting to all sections of the environment. As we have seen in this chapter, the scarcity of energy sources and the need to find alternatives has implications for air quality. How our resources should be used and what role pollution prevention should play in these decisions requires careful thought and long term planning. Environmental quality is one among the many goals a society has; a factor to be considered but not the only one.

Bibliography

Chapter 1   What is Air Pollution?

JAFFE, L. S., (1975) 'The Global Balance of Carbon Monoxide' in *The Changing Global Environment*, Ed. F. S. Singer, D. Reidel, Dordrecht.

LOWRY, W. P., (1967) 'The Climate of Cities', *Scientific American*, **217**, 15–24.

ROBINSON, E. and ROBBINS, R. C., (1972) 'Emissions, Concentrations, and Fate of Gaseous Atmospheric Pollutants' in *Air Pollution Control – Part II,* Ed. W. Strauss, Wiley Interscience, New York.

SINGER, F.S., (1975) 'Pollution Effects on Global Climate' in *The Changing Global Environment*, Ed. F. S. Singer, D. Reidel, Dordrecht.

STERN, A. C., WOHLERS, H. C., BOUBEL, R.W. and LOWRY, W. P., (1973) *Fundamentals of Air Pollution*, Academic Press, New York.

SVENSSON, B. H. and SODERLUND, R., Eds. (1975) *Nitrogen, Phosphorus and Sulphur – Global Cycles*, Swedish Natural Science Research Council, Stockholm.

Chapter 2   Sources of Air Pollution

BOND, R. G. and STRAUB, C. P., Eds. (1974) *Handbook of Environmental Control. Vol. 1 Air Pollution*, C.R.C. Press, Cleveland.

BUTLER, J. D., (1979) *Air Pollution Chemistry*, Academic Press, New York.

(1973) *Compilation of Air Pollutant Emission Factors*, 2nd Edition, US Environmental Protection Agency, Research Triangle Park, North Carolina.

PARKER, A., Ed. (1978) *Industrial Air Pollution Handbook*, McGraw-Hill, Maidenhead.

STERN, A. C., Ed. (1977) *Air Pollution Vol. IV*, 3rd Edition, Academic Press, New York.

STRAUSS, W., (1966) *Industrial Gas Cleaning*, Pergamon Press, New York.

Chapter 3   The Effects of Air Pollution

BAILEY, R. A., CLARKE, H. M., FERRIS, J. P., KRAUSE, S. and STRONG, R. L., (1978) *Chemistry of the Environment*, Academic Press, New York.

EICHOLZ, G. G., (1976) *Environmental Aspects of Nuclear Power*, Ann Arbor Science, Michigan.

FERRIS, B. G., (1978) 'Health Effects of Exposure to Low Levels of Regulated Air Pollutants', *J. Air Poll. Contr. Assoc.*, **28**, 482–97.

GLASS, N. R., GLASS, G. E. and RENNIE, P. J., (1979) 'Effects of Acid Precipitation', *Env. Sci. Tech.*, **13**, 1350–55.

HEICKLEN, J., (1976) *Atmospheric Chemistry*, Academic Press, New York.

JACOBSON, J. S. and HILL, A. C., Eds. (1970) *Recognition of Air Pollution Injury to Vegetation: A Pictorial Atlas*, Air Pollution Control Association, Pittsburgh.

KERR, J. A., CALVERT, J. G. and DEMERJIAN, K. L., (1972) 'The Mechanism of Photochemical Smog Formation', *Chemistry in Britain*, **8**, 252–7.

LAVE, L. B. and SESKIN, E. P., (1977) *Air Pollution and Human Health*, John Hopkins University Press, Baltimore.

PITTS, J. N., (1979) 'Keys to Photochemical Smog Control', *Env. Sci. Tech.*, **11**, 456–61.

STERN, A. C., Ed. (1977) *Air Pollution Vol. II*, 3rd Edition, Academic Press, New York.

THRUSH, B.A., (1979) 'Aspects of the Chemistry of Ozone Depletion', *Phil. Trans. R. Soc. Lond.*, A **290**, 505–14.

WOODWELL, G. M., (1978) 'The Carbon Dioxide Question', *Scientific American*, **238**, 34–43.

Chapter 4   Measurement of Pollutants

BRENCHLEY, D. L., TURLEY, C. D. and YARMAC, R. F., (1974) *Industrial Source Sampling*, Ann Arbor Science, Michigan.

CADLE, R. D., (1975) *The Measurement of Airborne Particles*, Wiley Interscience, New York.

PERRY, R. and YOUNG, R., Eds. (1977) *Handbook of Air Pollution Analysis*, Chapman and Hall, London.

SCHNEIDER, T., Ed. (1974) *Automatic Air Quality Monitoring Systems*, Elsevier, Amsterdam.

STERN, A. C., (1977) *Air Pollution Vol. III*, 3rd Edition, Academic Press, New York.

STRAUSS, W., Ed. (1979) *Air Pollution Control – Part III*, Wiley Interscience, New York.

WARNER, P. O., (1976) *Analysis of Air Pollutants*, Wiley Interscience, New York.

Chapter 5   Air Pollution Control

PARKER, A., Ed. (1978) *Industrial Air Pollution Handbook,* McGraw-Hill, Maidenhead.

STERN, A. C., Ed. (1976) *Air Pollution Vol. IV,* 3rd Edition, Academic Press, New York.

STORCH, O., (1979) *Industrial Separators for Gas Cleaning,* Chemical Eng. Monographs 6, Elsevier, Amsterdam.

STRAUSS, W., (1966) *Industrial Gas Cleaning,* Pergamon Press, New York.

WALTERS, J. K. and WINT, A., Eds. (1981) *Industrial Effluent Treatment. Vol. 2 Air and Noise,* Applied Science Publishers, London.

Chapter 6   Non-Technical Aspects of Control

FREEMAN, A. M., HAVEMAN, R. H. and KNEESE, A.V., (1973) *The Economics of Environmental Policy,* John Wiley and Sons, New York.

HOLDGATE, M. W., (1979) *A Perspective of Environmental Pollution,* Cambridge University Press, Cambridge.

KNEESE, A. V., (1977) *Economics and the Environment,* Penguin Books.

LANTERI, A., (1970) 'The Philosophy of Air Pollution Legislation', *Clean Air,* **4,** 45–46.

STERN, A. C., Ed. (1976) *Air Pollution Vol. V,* 3rd Edition, Academic Press, New York.

Chapter 7   The Future

BAILEY, R. A., CLARK, H. M., FERRIS, J. P., KRAUSE, S. and STRONG, R. L., (1978) *Chemistry of the Environment,* Academic Press, New York.

HODGES, L., (1977) *Environmental Pollution,* 2nd Edition, Holt, Rinehart and Winston.

MEETHAM, A. R., (1981) *Atmospheric Pollution: Its History, Origins and Prevention,* 4th Edition, Pergamon Press, Oxford.

Index

Acid digestion, 77
Acidification of natural waters, 69, 70
Acid mists, 95, 96
Acid rain, 69, 70, 129, 139
Acropolis, 62
Activated carbon, 91, 92, 108
Acute effects, 6, 49, 50
Adiabatic lapse rate, 8, 103, 104;
    see also Lapse rate
Adirondaks, USA, 70
Adsorbents, 78, 91, 92
Adsorption of gases, 91, 92
Aerodynamic properties, 77
Aerosols, see Particulates
Air: composition of, 1, 2
Aircraft emissions, 33, 34
Air pollution:
    criteria, 36, 71, 112, 113, 115, 121
    episodes, 7, 37
    exposure, 113, 114
    Management Authorities, 119, 121
    models, 65–7, 85, 115, 117
    monitoring, 83, 85, 122
Air quality standards, 72, 112, 114,
    115, 117, 119
Airshed:
    management, 116
    models, see Air pollution models
Air to fuel ratios, 33, 110, 111
Albedo, 66
Alcohol vehicles, 140
Aldehydes, 33, 34, 41–7
Alkali Act, UK, 7, 20, 55, 120
Alkali and Clean Air Inspectorate, 120
Alpha radiation, 51, 52
Alumina, 91, 128
Aluminium industry, 7, 61, 128
Alveoli, 47
Amine treatment, 89, 91
Ammonia, 2, 3, 18, 24, 25, 91
Ammonium sulphate fertilizer, see
    Fertilizers
Amsterdam, 127
Anaconda smelter, 61
Anaemia, 50
Animals, effect of pollutants on, 60–62

Anti-knock additives, 31, 34, 109
Aromatic hydrocarbons, 93, 128
Arsenic, 16, 61, 105
Art works: effect of pollutants on, 37, 62
Asbestos, 22, 47
Asbestosis, 47
Ash content of coal, 26, 27, 87, 88, 107, 108
Asthma, 39, 48, 49
Athens, 62
Atomic absorption, 77
Atomic fission products, 52
Atomic oxygen, 42, 67
Auckland, New Zealand, 62
Australia, 13, 35, 120, 121, 123, 126
Automobile:
    emissions, 7, 12, 29–35, 39, 74, 75,
        108–11, 118, 121, 122, 139
    emission limits, 109, 121, 122
Automative distillate, see Diesel fuel

BWR, 133, 135
Background concentrations, 1, 2, 36, 72
Bacterial action, 3
Bag houses, 100, 101
Bailey, R.A., 144, 145
Barium additives to fuel, 111
Batteries, 130, 140
Behavioural problems in children, 50
Belton, 114
Benchley, D.L., 144
Benzo(a)pyrene, 29, 31, 33, 34
Best practicable means, 116, 119
Beta radiation, 51, 52
Biological response time, 72, 115
Black liquor, 22, 23, 28, 93
Blast furnace, 18, 19
Blood lead levels, 49, 50
Boiling water reactor, see BWR
Bond, R.G., 29, 33, 143
Boubel, R.W., 99, 104, 141
Brazil, 140
Breeder reactors, 134, 135
Bronchial tubes, 47
Brown coal, 22
Brownian motion, 99
Buffer zones, see Green belts

Building materials: effect of pollutants on, 37, 62
Buoyancy lift of plumes, 105
Burnt lime, 23
1–4 Butadiene, 42
Butler, J.D., 33, 81, 143

CANDU reactors, 133
Cadle, R.D., 144
Cadmium:
  animal health effects, 61
  human health effects, 50
  measurement of, 77
Calibration standards, 85
California, 121
California cycle, 109
Californian Rule 66, 42
Canada, 9, 12, 118, 126, 131, 133
Canadian heavy water reactor, *see* CANDU
Cancer, 50–55
Carbon black, 13
Carbon: emissions of, 27, 34
  in iron and steel, 19
  reducing agent, 16, 18
Carbon dioxide:
  general, 2, 3, 16, 25, 73, 118, 128, 129
  and earth's radiation balance, 5, 6, 64–7
Carbon monoxide:
  general, 2, 19
  health effects, 49
  measurement, 73, 80, 81, 84
  removal, 92–4
  sources, 3, 7, 18, 19, 25–34, 108–11, 121, 122, 128
Carboxyhaemoglobin, 49
Carburettor, 30, 31, 108
Carcinogens, 50, 51; *see also* Individual carcinogens
Car pooling, 141
Cassava, 140
Catalyst poisoning, 18, 92, 105, 109
Catalytic cracking, 13–15
Catalytic:
  gas cleaning, 92, 94, 108–11
  oxidation, 15, 17, 38, 94, 105, 106, 109, 111
Cement kilns, 22, 96
Centrifugal spray scrubber, 101
Cerebral palsy, 50
Chain reactions in photochemical smog formation, 42–5
Chattanooga, 48, 49
Chemical plants, 12, 20, 21, 28, 29
Chemiluminescence nitric oxide analyser, 83

Chemiluminescence ozone monitor, 83
Chemisorption, 78, 79, 92
Chemisorption tubes, 79
Chicago, 34
Chimney design, 104
Chlorofluoromethanes, 5, 6, 36, 68, 69, 118
Christchurch, New Zealand, 25
Chromatography, 77; *see also* Gas, Thin layer and Liquid chromatography
Chromosome damage, 53
Chronic bronchitis, 39
Chronic effects, 6, 37, 39, 48–55, 112
Cigarette smoking, 39, 49, 51, 113
City climate, 9, 10
'Classical' smog, *see* London smog
Claus process, 15
Clean Air Acts, UK, 120
Clean air zones, 88
Climatic change, 65–7, 129
Coal:
  conversion, 139
  gasification, 25, 126
  heat value, 25, 29
  rank, 26
  reserves, 25
Coal combustion:
  commercial or domestic boilers, 34, 125, 126
  electricity generation, 25–7, 34, 106–8, 130
  emissions, 38, 39, 51
Coalite, 126
Co-current scrubbing, 90
Coefficient of haze, 76
Coke, 19, 126
Coke oven odours, 19
Cologne, 62
Combustion modification for emissions control, 87, 107, 108, 110, 111, 139
Combustible waste gases, 92–4
Comparative source strengths, 34
Composting, 25
Compression ratios, 109
Coning, 103, 104
Conservation of energy, 127, 140, 141
Controlled sources, 122
Control strategies, 116, 117
Copper, 128
Copper smelting:
  general, 15–17, 105
  vegetation damage from, 15, 16; *see also* Non-ferrous metal smelting
Core meltdown, 137
Costs:
  of air pollution damage, 62, 63, 115

of air pollution control, 95, 104, 105, 116
of energy alternatives, 127, 131, 132
Counter current flow, 88
Coventry, 117, 120
Crankcase emissions, 31, 108
Crowley, D., 114
Cryogenic gas collection, 78
Cumulative effects, 72
Cyclones, 98, 99, 131

Deashing of coal, 87
Denver, 40
Desulphurization, 87, 106
Deuterated water, 133
Deuterium, 138, 139
Diatomic sulphur, 81
Diesel engine:
    general, 30–33, 110, 111
    emissions, 31–4, 139
Diesel fuel, 13, 14
Dimethylsulphide: measurement of, 82
Di-radicals, 42–5
District heating, 28, 127
Domestic heating, 25, 125–7
Donora, Pennsylvania, 61
Dow Chemical Company, 29
Dowtherm, 29
Driving mode, 31, 109, 123
Dublin, 114
Ducktown, Tennessee, 58
Dust, 120
Dust collectors, 94–102
Dustfall, *see* Particle fallout
Dynamometer, 74

Economic incentives, 123
Effective stack height, 103, 104
Eicholz, G.G., 136, 144
Einsteins law, 132
Electric heating, 126
Electricity generation:
    general, 12, 13, 25–9, 106–8, 128, 139,
        140
    emissions from, 29, 34, 35, 106–8,
        128–139
Electrostatic precipitation, 95–8, 108
Emission controls, 116, 117, 119, 120
Emissions inventory, 123
Emphysema, 39, 61
Enrichment of uranium, 136
Environmental impact statements, 123
Environmental Protection Agencies, *see*
    Air Pollution Management
    Authorities
Enzyme reactions in plants, 59
Epidemiological studies, 37, 48

Ethanol, 140
Europe, 9, 12, 69, 70, 104, 118, 120, 125,
    138
Exposure factor and plant damage, 57
External combustion, 25–9
Eye irritation, 46

FPD method for sulphur, 82
Fabric filters, 95, 99–101
Fanning, 103, 104
Federal Clean Air Legislation, USA, 120
Federal and State jurisdiction in air
    pollution control, 120, 121
Feedback mechanisms, 65
Ferris, B.G., 141
Fertilizer manufacture:
    from composting, 25
    from waste gases, 18, 89, 105
    general, 7, 13, 20, 21, 61
Filters, 76, 95, 99–101
Finland, 50
Fish mortality, 70
Fission reactions, 132, 134
Flame ionization detector, 83, 84
Flame photometric detector, 82
Flexible working hours, 141
Flowers: pollutants and, 56
Fluorides:
    effects on animals, 61
    effects on plants, 60
Fluorescence sulphur dioxide analyser, 82
Fluorescent spectroscopy, 77
Fluoride emissions, *see* Hydrogen fluoride
Fluorosis, 61
Fluorsilicic acid, 21
Fly ash, 26–8, 34, 94, 96, 106, 108
Fog, 10, 25
Food:
    manufacture, 24
    odours, 24, 94
Forest destruction, 3, 64–7
Formaldehyde, 42, 44, 45
Fossil fuels, 64, 65, 125, 127, 137, 139;
    see also Coal and Oil combustion
Foundries, 19
Foundry odours, 19
Freeman, A.M., 145
Fuel cells, 131, 140
Fuel:
    combustion, 7, 9, 25–34; *see also*
        Individual fuels
    heating values, 29
Fuel oil:
    for electricity generation, 106
    general, 13, 14, 126
Fuel rods, 133, 138

Fuel tank emissions, 31, 108
Fugitive dust, 19, 22
Fumigation, 103, 104
Fungicides, 60

Gamma radiation, 51, 52
Garbage incineration for district heating, 127
Gas absorption in liquids, *see* Scrubbing
Gas bubblers, 78
Gas chromatography, 77, 82
Gas diffusion for uranium enrichment, 133
Gas fired boilers:
  commercial and domestic, 28, 34, 126
  electricity generation, 25, 28
Gas oil, 13, 14
Gasoline, *see* Motor spirit
Gas sampling, 83
Genetic effects, 53
Geothermal energy, 128, 130, 131
Germany, 28, 61, 85, 110, 126
Glass, N.R., 144
Glass manufacture, 22
Gothic cathedrals: air pollution damage to, 62
Grab sampling, 77, 85
Gravity separation, 95, 96
Gravity settling chamber, 95
Green belts, 88, 118, 123
Greenhouse effect, 64–7
Griess–Saltzmann methods, 79
Grit, 120

Haze, 10
Heart disease, 49, 50, 61
Heat island effect, 9, 10
Heat loss and recovery in power stations, 27, 28
Heat storage:
  in liquids, 127
  in rock piles, 127
  in Sodium Sulphate, 127
Heavy metals:
  animal health effects, 61
  effects on plants, 60
  human health effects, 47, 49, 50
Heavy water, *see* Deuterated water
Heicklen, 144
Helium, 133
Helium nucleii, 51, 52
High volume sampler, 76, 77
High volume shelter, 76
Hill, A.C., 58, 144
Hiroshima, 53
Hodges, L., 54, 145
Holdgate, M.W., 88, 145

Humidity, 47, 77
Hydrocarbon reactivity scale, 42
Hydrocarbons:
  as petrochemical feedstock, 21
  from organic solvents, 23
  measurement of, 83
  removal from waste gas streams, 92–4
  role in smog formation, 39–47
  sources of, 3, 7, 15, 25, 29, 31, 33, 34, 108–11, 121; 122
  see also Methane and Non-methane hydrocarbons
Hydrochloric acid, 2, 7, 91, 95
Hydrodesulphurization, 13, 14
Hydroelectricity, 130–32
Hydrogen fluoride, 2, 7, 18, 21, 60, 61
Hydrogen peroxide method for SO_2, 79
Hydrogen sulphide:
  conversion to SO_2, 3
  effect on materials, 62
  measurement of, 82
  removal from waste gases, 89, 91
  sources of, 3, 7, 15, 22, 23, 128
Hydroxyl radical, 43–5, 67, 68

Ice ages, 66
Iceland, 128, 130
Incineration:
  general, 12, 24, 25
  odours, 24
Incomplete combustion, 88
Industrial accidents, 37
Industrial boilers, 13, 28
Inert gases, 63
Infra red absorption, 6, 64–7
Infra red spectroscopy, 73
Inorganic metal content of fuels, 12, 34
Insecticides, 60
Insulation, 127
Internal combustion engine, 29–34; *see also* Spark ignition and Diesel engine
International problems, 118, 119
Inversion, 8, 10, 39, 40
Iodine 131, 137
Ionizing radiation, *see* Radioactivity
Iron:
  as a catalyst in smog formation, 38
  manufacture of, 18, 19
Irradiation times, 42
Iso-kinetic sampling, 74
Itai-itai disease, 50
Italy, 37

Jacobson, J.S., 58, 144
Jaffe, L.S., 143

Japan, 75, 85, 120
Jet planes, 33
Johannesburg, 9

Keeling, Charles, D., 65
Kerosene, 13, 14, 19
Kerr, J.A., 144
Kew, England, 88
Kneese, A.V., 145
Kraft process, 22, 23
Krypton, 85, 137

LPG, 109, 140
LWR, 133; *see also* PWR and BWR
Lake Manapouri, 132
Lake Pedder, 132
Landfill, 24, 25
Land use planning, 118, 141, 142
Lanteri, A., 145
Lapse rate, 7–9, 103–5
Larderello, Italy, 130
Lave, L.B., 144
Lead:
    animal health effects, 61
    automobile emissions, 31, 34, 49, 109,
        110
    general, 5, 18, 77, 92
    human health effects, 49, 50
Lead peroxide candle, 78, 79
Lead smelting, 15, 16, 105, 128; *see also*
    Non-ferrous metal smelting
Leaf structure, 55
Lean fuel mixtures and emissions, 31, 109
Learning problems in children, 50
Legislation: air pollution control, 119, 120
    USA, 120
    UK, 120
    Australia, 121
Legislative control, 88
Leonard, A.G., 114
Leukaemia, 53
Light water reactors, *see* LWR
Lignite, *see* Brown coal
Lime slurries, 91, 106
Limestone, 17, 22, 106
Liquid chromatography, 77
Liquid fuel from crops, 140
Liquid metal cooled fast breeder reactor,
    135
Liquid sodium coolant, 135
Local government responsibility, 120
Lofting, 103, 104
London, 8, 28, 38, 61, 88, 117, 120, 127
London smog, 7, 8, 10, 38, 61, 125
Looping, 103, 104
Los Angeles, 8, 10, 39, 41, 46, 62, 86, 109

Los Angeles cycle, 109
Low level emissions, 35
Lowry, W.D., 99, 104, 143
Lung, 47–55
Lung cancer, 50, 51, 53, 136
Lurgi process, 126

Magnesium: as a catalyst in smog
    formations, 38
Magnesium/manganese oxide
    slurries, 106
Magneto-hydrodynamics (MHD), 131
Marshall Islands, 53
Mass spectroscopy, 77
Mauna Loa, Hawaii, 64, 65
Maximum allowable concentrations, 48
Meetham, A.R., 145
Memorandum of Intent, 118
Mental retardation, 50
Mercaptans, 7, 19, 22–4, 82
Metal smelting, *see* Non-ferrous metal
    smelting
Methane, 2, 19, 40, 84, 140
Methanol, 140
Mexico, 130
Mexico City, 9
Milling uranium ore, 134, 135
Mineral content of fuels, 27–9, 34, 35;
    see also Ash content of coal
Mineral extraction, 22
Mining, 22, 135, 136
Mixing depth, 8
Monitoring networks, 85, 86
Monitoring station location, 122
Mortality:
    animal, 61
    human, 37, 38, 46, 49
Motor cars, *see* Automobiles
Motor spirit, 13, 30, 35
Mount Isa, Queensland, 18

Nagasaki, 53
Natural emissions of gases, 3, 4
Natural gas, *see* Gas
Netherlands, 75, 85
Neutral sulphide semi-chemical process,
    22, 23
Neutron bombardment, 138
Neutrons, 133, 135, 138, 139
Newcastle, Australia, 62
New Guinea, 50
New York, 28, 50, 127
New Zealand, 62, 128, 131
Nitric acid, 38
Nitric oxide, 2, 68, 83; *see also* Nitrogen
    oxides

Nitrogen dioxide, 2, 59, 83; *see also*
    Nitrogen oxides
Nitrogen oxides:
    control in waste gases, 92
    health effects, 48, 49
    in ozone destruction, 68
    in smog formation, 11, 38–46
    measurement of, 78, 83
Nitrous acid, 45
Nitrous oxide, 2, 3, 67, 68
Non-dispersive infra red gas analyser, 81
Non-ferrous metal smelting, 4, 12, 15–18,
    55, 60, 61, 105
Non-methane hydrocarbons, 2, 3, 40–47,
    84
Northern Hemisphere, 4
Norway, 131
Nuclear accidents: probability of, 137, 138
Nuclear fusion, 138, 139
Nuclear power plants, 25, 29, 54, 55, 129,
    132–9
Nuclear reactors, 132–9; *see also* BWR,
    PWR, CANDU

OECD Study of acidification, 69
Occupational exposure, 37
Ocean circulation, 64, 66
Oceans as a source of deuterium, 139
Ocean temperature gradient for power
    supply, 130
Octane number, 109
Odour, 55, 91, 93, 94
Off peak electricity, 126
Oil combustion: for electricity generation,
    25, 28, 34
Oil fired boilers: commercial and
    domestic, 28, 34
Oil refineries, 12, 21, 28
Oil shale, 139
Olefines, 42
Open cut mining, 22
Open fires, 25, 126
Organic acids, 33, 34
Organic amines, 84
Organic gases: collection of, 78
Organic solvents, 23, 24
Organic sulphides, 93; *see also*
    Mercaptans and Dimethylsulphide
Orsat method, 73
Otto engine, *see* Spark ignition engine
Oxidant:
    general, 40–47
    measurement of, 80, 83
Oxygen, 2, 49, 73
Ozone:
    atmospheric concentration, 2

destruction in the stratosphere,
    67–9
effect on plants, 58, 59
effect on materials, 62
measurement of, 83
photochemical smog production, 40–47

PAN, *see* Peroxyacetyl nitrate
PWR, 133, 134
Packed absorption tower, 89, 90
Paint:
    emissions from, 23, 24
    manufacture, 20
    odours, 23, 24, 94
Palisade cells, 56, 59
Paper pulp manufacture, 22, 23, 28, 93
Parker, A., 74, 98, 102, 143, 145
Particle composition, 77
Particle fallout, 75
Particles:
    changing concentration, 66
    effects on health, 114
    effects on materials, 62
    measurement of, 73
    radiation balance, 66
    smog formation, 10, 38, 39, 46, 47
    sources of, 5–7, 15, 19, 23, 29–34,
        106–8, 128
Particle size, 46, 77, 95, 96
Particulate metals, 77
Particulate organics, 77
Peak demand for electricity, 132
Peak hour traffic, 35
Peroxyacetyl nitrate, 40, 43, 59, 60
Peroxyhydroxyl radical, 39–47, 67, 68
Perry, R., 84, 144
Petrochemical industry, 13, 21
Petrol, *see* Motor spirit
Petroleum refining, 12–15, 35, 91
Philadelphia, 40
Phillips, A., 20
Phosphate rock, 18, 21
Photochemical reactions, 39–47, 67, 68
Photochemical smog:
    general, 10, 39–47, 85, 86, 107
    hydrocarbon precursors, 33
    particulate formation, 33
    solvent contribution, 23
Photosynthesis, 59, 127, 128
Photovoltaic effect, 130
Pitch, 19
Pitts, J.N., 144
Pittsburgh, 125
Plants: effect of pollutants on, 37, 55–60
Plant indicators, 57
Plant sensitivities, 57–60

Plastic bags, 77
Plastics, 13, 20
Plate electrostatic precipitator, 97
Plutonium 239, 133–5, 138
Polar ice caps, 66
Polyaromatic hydrocarbons, 6, 29, 33, 51,
    77, 128; *see also* Benz(a)pyrene
Ponderosa Pine, 59
Potassium iodide method for oxidant, 80
Power stations, *see* Electricity generation
Pratt, G., 16
Pressurised water reactor, *see* PWR
Psychological effects of odours, 24
Public education, 124
Public transport, 124, 141, 142
Pumped storage, 132
Pyrites, 21
Pyrosynthesis, 33

Quarrying, 22
Queenstown, Tasmania, 16, 58

Rad, 52
Radiation dosage, 52–5
Radical species, 42–7
Radioactive sodium, 135
Radioactive waste, 132, 136–8
Radioactivity: health effects of, 51–5, 136
Radionuclides, 137, 138; *see also*
    Individual isotopes
Radon, 128, 136
Rance estuary, 130
Rainbow Bridge, Utah, 132
Real time monitoring, 72, 75
Recycling, 25
Reflectance units of dirt shade, 76
Reforming, 13, 14
Rem, 52
Rendering, 24
Rendering odours, 84
Respiratory problems, 38, 39, 46, 47, 76,
    114
Reverberatory furnace, 17
Rich fuel mixtures, 31
Rijmond, Netherlands, 86
Ringleman chart, 74
Ringleman number, 73
Robbins, R.C., 143
Robinson, E., 143
Roentgen, 52
Rotorua, New Zealand, 62
Rotterdam, 46
Rubber, 62

Safety precautions in nuclear power
    plants, 137

Sampling: representativeness of, 74
Sand, *see* Silica
Sawdust, 98
Scandinavia, 67, 69, 118
Schneider, T., 144
Scrap metal, 19
Scrubbing of gases, 89–91, 106
Scrubbing of particles, 101, 102
Self monitoring, 121
Sellafield, 137
Seveso, 37
Silica, 17, 19, 22, 47
Silica gel, 91
Silicon semiconductors, 130
Silicosis, 47
Silverware: pollution tarnishing, 62
Singer, F.S., 141
Skin cancer, 69
Smog: London, *see* London smog
Smog: photochemical, *see* Photochemical
    smog
Smoke, 27, 28, 38, 39, 46, 88, 94, 120
Smoke consuming appliances, 88, 89
Smoke density measurement, 73
Smokeless fuels, 88, 89, 126
Smokeless zone regulations, 39
Soderlund, R., 141
Sodium carbonate, 23
Sodium hydroxide, 22, 23
Sodium sulphate, 127
Sodium sulphide, 22, 23
Sodium sulphite, 91, 106
Soil:
    effect of acidity on, 69, 70
    transfer of pollutants to, 55
Solar electricity, 129–130
Solar energy, 127–130
Solar home heating, 127
Solar hot water, 127
Solid adsorbents for gas sampling, 78
Solvent recovery, 92
Somatic effects, 53
Soot, 111
Sore throats, 46
Source monitoring, 72–5
South Africa, 126
Soviet Union, 130
Space heating, 12
Spark ignition engine, 30–34
Special air quality zones, 118
Stability, atmospheric, 7, 8, 38, 39, 103–5
Stack gas analysis, 73
Stack monitoring, *see* Source monitoring
Stacks, 74
Steam bores, 130, 131
Steam generator, 133

Steel: manufacture of, 18, 19
Stern, A.C., 17, 23, 33, 99, 100, 104, 109, 141, 144, 145
Steubenville, Ohio, 63
Stockholm, 127
Stokes Law, 95
Stomata, 56
Storch, O., 145
Stratified charge engine, 111
Stratosphere, 5, 34, 63, 64
Stratospheric ozone, 67–9
Straub, C.P., 29, 33, 143
Strauss, W., 5, 82, 86, 90, 93, 95–7, 101, 103, 134, 135, 143–5
Subadiabatic lapse rate, 9
Sugar, 140
Sulphur:
   in fuels, 12–15, 20, 26–9, 34, 86, 106, 108
   in minerals, 15, 18, 19, 105
   removal from fuels, 13, 142
Sulphur dioxide:
   atmospheric concentration, 2, 9, 39
   effects on health, 38, 39, 114
   effects on plants, 55–8
   measurement of, 78, 79, 81, 82
   role in smog formation, 10, 38, 39, 46
   sources of, 3–7, 15–17, 20, 21, 33, 34, 93, 105–8, 128
   sulphuric acid manufacture from, 18, 21
   transport of, 69, 70, 118, 119, 129
Sulphur trioxide, 39, 93, 105, 108
Sulphuric acid:
   catalyst, 21
   from atmospheric SO_2 oxidation, 6, 38, 46, 105
   manufacture, 15, 18, 20, 21, 102, 105
   removal from waste gas streams, 90, 91, 95
Super phosphate, *see* Fertilizers
Supersonic aircraft emissions, 34
Surface area of adsorbents, 91
Svenson, B.H., 141
Sweden, 28, 61, 69, 70
Switzerland, 131
Sydney, 46
Synergism, 7, 36, 38, 39, 46, 51, 56, 59, 112, 114
Synthetic fibres, 100, 101
Synthetic rubber, 13

TCDD, 37
TLV, *see* Threshold limiting value
Tall stacks, 18, 35, 103–5
Tangential firing, 107
Termination reactions, 45

Thames Valley, 8, 106
Thermal conductivity device, 73
Thermal power stations, 25–9, 106–8, 128–132; *see also* Coal and Oil combustion
Thin layer chromatography, 77
Thompson, R., 20
Thorium 232, 135
Three Mile Island, Pennsylvania, 137
Threshold limiting value, 39, 48
Thrush, B.A., 144
Tidal power, 130
Tobacco plants, 59
Tokyo, 46
Town gas, 19
Toxic emissions, 118
Train transport, 141, 142
Tritium, 137–9
Troposphere, 1, 69
Turbines, 27, 132, 133

Ultra violet absorption, 65, 67
Ultra violet radiation, 6
Ultra violet spectroscopy, 73
Uncontrolled sources, 122
United Kingdom, 69, 79, 120, 126, 132, 137, 138
United States Environmental Protection Agency, 33, 40, 41, 63, 129
United States National Academy of Sciences, 54
United States of America, 9, 12, 35, 55, 70, 75, 104, 109, 118, 120, 122, 125, 126, 129, 132, 137
United States Regulatory Commission Report, 137
Uniontown, Pennsylvania, 62
Uranium, 132, 139; *see also* Individual isotopes
Uranium 233, 134
Uranium 235, 132
Uranium 238, 133–5
Uranium hexafluoride, 133
Uranium oxide, 133
Urban factor: in lung cancer, 50, 51
Urban planning, 88

Vanadium: as a catalyst in smog formation, 38
Vapours, 5, 6
Velocity rise of plume, 104
Venturi scrubber, 101, 102
Visibility, 5, 46, 47, 140
Volcanoes, 3, 4, 67

WHO, 50

WMO, 69
Wairakei, New Zealand, 131
Walters, J.K., 145
Warner, P.O., 144
Washington, 40
Water based paints, 23
Water pollution from landfill, 25
Wave power, 129, 130
Wawa, Ontario, 58
Weedicides, 60
West Gaeke method, 79
Wind power, 129, 130
Wire in tube precipitator, 96
Wohlers, H.C., 99, 104, 141
Wood combustion, 12, 25, 128

Wood pulping industry, 7, 22, 23
Woodwell, G.M., 144
World War II, 114

X-rays, 51, 52

Young, R., 84, 144

Zinc:
    general, 18, 60, 77
    smelting, 15, 16, 105, 128; *see also*
        Non-ferrous metal smelting
Zirconium alloy tubes, 133
Zoning of industry, 87
Zürich, Switzerland, 88